Land Alienation and its Dimensions

The Institute of Social Sciences

The Institute of Social Sciences was founded in 1985, to study contemporary social, political and economic issues and problems in an inter-disciplinary perspective and to make available its findings and recommendations to government bodies, social scientists, policy makers, people's and workers' organisations, so as to widen their options for action. The evolution of an informed and action-oriented public opinion is the primary aim of the Institute.

The research projects undertaken by the Institute cover a wide range of subjects in the areas of local governance, women's studies, environment and contemporary economic and political issues. The Institute also organises seminars, workshops, discussions and training programmes for the exchange of ideas and dissemination of its research findings.

The major research thrust of the Institute is in the areas of Local Governance (*Panchayati Raj*), Urban Studies, Women's Studies, Economic Affairs, Human Rights and Police Reforms and International Studies.

We seek to build a community of concerned scholars and activists engaged in ushering in a humane and just society.

About the Authors

Shri Karunakar Patnaik is a former civil servant who retired as Director of Panchayati Raj Department, Government of Orissa. With long experience spanning over thirty-five years in public service, he specialized both in rural and urban governance. Presently, he is the regional head of the Institute of Social Sciences, Bhubaneswar.

Dr. Pramila Prava Patnaik previously was working as Lecturer in Economics in different colleges of Orissa for 12 years. She was consultant in Economics in State Institute of Rural Development for 2 years. Presently, she is working as Research Officer in Institute of Social Sciences, Bhubaneswar branch. She has authored two research study books. Her research papers and articles are published in books and journals.

Land Alienation and its Dimensions

A Study of Scheduled Districts of Orissa

Karunakar Patnaik
Pramila Prava Patnaik

Institute of Social Sciences
Eastern Regional Branch, Bhubaneswar

CONCEPT PUBLISHING COMPANY PVT. LTD.,
NEW DELHI-110059

ISBN-13: 978-81-8069-766-1

First Published 2011

Published and Printed by

Concept Since 1974

Concept Publishing Company Pvt. Ltd.
Regd. Office:
A/15-16, Commercial Block, Mohan Garden,
New Delhi-110059 (INDIA)
Phones : 25351460, 25351794, *Fax* : 091-11-25357109
Email : publishing@conceptpub.com,
Website: www.conceptpub.com

Editorial Office:
H-13, Bali Nagar, New Delhi-110 015, India.

Cataloging in Publication Data--*Courtesy:* D.K. Agencies (P) Ltd. <docinfo@dkagencies.com>

Patnaik, Karunakar.
Land alienation and its dimensions : a study of scheduled districts of Orissa / Karunakar Patnaik, Pramila Prava Patnaik.
p. cm.
Includes bibliographical references (p.).
ISBN 9788180697661

1. Allotment of land--India--Koraput (District) 2. Allotment of land--India--Rayagada (District) 3. Land tenure--India--Koraput (District) 4. Land tenure--India--Rayagada (District) 5. Koraput (India : District)--Scheduled tribes--Economic conditions. 6. Rayagada (India : District)--Scheduled tribes--Economic conditions. 7. Rural development--India--Koraput (District) 8. Rural development--India--Rayagada (District) I. Patnaik, Pramila Prava, joint author. II. Title.

DDC 333.10954 22

Foreword

Land is the most precious asset for human beings, particularly for the poor. Since the value of land is ingrained deep in the consciousness of humanity in terms of their well-being, it provides the base for their identity. Land has also become the symbol of social status. It is, in fact, the major source of livelihood for the large majority of humankind everywhere especially the tribals. Directly and indirectly, land affects their livelihood options.

In India, industrialization, mining operations and 'development' projects have dispossessed large sections of tribal people and taken away, their source of livelihood. They have been dispossessed of their homes and hearths due to illegal and fraudulent transfer of lands. Despite legal protection and enforcement of multiple rules and regulations, land alienation has become the norm of the day displacing *adivasis, dalits* and other marginalized people of their entitlement to land resulting in chronic indebtedness and placing them in precarious conditions.

Orissa's poverty level is one of the highest in India. The tribals bear the brunt of poverty and misery who constitute 22% of the state population. Today they have hardly any control over the forest land which is their habitation. Distribution of land has been highly unfair and unequal. Majority of people do not have access to land due to historical and political reasons and complicated land governance system in the state. It is against this background, the Institute of Social Sciences, through its regional branch in Bhubaneswar, conducted a study in the two districts of Koraput and Rayagada which come under Schedule V area. The overall objective of the study was to empower the tribals and deprived sections through awareness and consultation camps for realization of their rights and entitlements over land.

The operational area of the project was confined to three blocks each of Koraput and Rayagada districts. From each block two Gram Panchayats were selected covering 173 villages and 10130 households for intensive study. This was a stupendous task. This publication is the result of hard work and meticulous research for three years including field action. The research included baseline survey, Focus Group Discussions (FGD), direct interaction with the key-stakeholders, selection of model villages, informal interviews with key stakeholders like Panchayati Raj Institutions (PRI) representatives, officials, Non-Governmental Organisations (NGOs) / Community Based Organisations (CBOs), advocates and other cross-sections of society. The secondary data was collected from Government departments, published documents and records.

It is heartening to note that this project through its research study and action has created awareness among such a large number of tribals about various land related issues, Acts, laws and other legal provisions. The training and awareness building programmes through a campaign mode was extremely useful. It may also be noted that the project intervention included empowerment of target groups like Panchayati Raj institutions members, government functionaries, youths and women. This was an important milestone in the implementation of this project. The study covering review of land laws and regulations with specific reference to alienation, the impact of land legislations in relation to alienation, transfer and displacement and restoration of illegally transferred tribal lands is a commendable aspect of this work. The research team deserves all credit. The research findings have put in place several key results of the study in the two study districts. Above all, several gaps in the existing system of land issues and related legislations have been indicated which are very relevant as correctives for the administrators and law-makers. The coverage of more than twenty seven thousand people directly or indirectly in both the districts is remarkable. I am sure they could become change agents which is not a small achievement.

I am grateful to the Irish Embassy for sponsoring this project. The Irish Aid support came at a time when the issues discussed above were brought to the centre stage by thinkers and activists. From the beginning of our work in Orissa Mr. Karunakar Patnaik has shown extraordinary interest in the issues affecting the poor and the marginalized in the state. Mr. Patnaik, Senior Fellow and Ms. Pramila P. Patnaik, Research Officer, the Institute of Social Sciences, Eastern Regional Centre have taken our research and action forward in critical areas.

In view of the significance of the study and interesting findings it generated, it was decided to bring out this publication for wider dissemination. I am sure it will provide valuable resource material for all those who are concerned with land reforms and tribals particularly in Orissa.

New Delhi
14 November 2010

George Mathew
Director
Institute of Social Sciences

Preface and Acknowledgements

The separate province of Orissa was carved-out from the erstwhile province of Bihar and Orissa on the 31st March, 1936. Prior to this, different districts were included in Bengal Presidency, the central provinces and the Madras Presidency. On the 1st January 1948, 25 princely states (twenty three belonging to the Orissa State Agency and the remaining two to Chhatisgarh Agency) were merged in Orissa. Two of these namely Saraikella and Kharswan (now in Jharkhand) were transferred to Bihar. The princely states of Mayurbhanj merged in Orissa on the 1st January, 1949. With this, the state of Orissa was composed of 13 districts. Orissa has now 30 revenue districts. Because of association of different parts of the state with different provinces then, different revenue laws were in operation in various parts of the state. This diversity was further aggravated due to the necessity of special provision for Scheduled Castes and Scheduled Tribes of the state. Each of the princely states had its own system of revenue law based on either the pattern of revenue administration prevalent in the then Bengal, or the Madras pattern of revenue administration. In all, as many as 32 different laws and administrative system governing land and rights therein were in vogue. These facts made the land revenue administration of the state very complicated, perhaps one of the most complicated in India.

The undivided Koraput district presently consisting of Koraput, Rayagada, Malkangiri and Nawarangpur district was established as a separate district of Odisha on the formation of separate Odisha state in 1936 with Mr. R.C.S. Bell, ICS as its first Collector. Land administration was the survival of the ancient feudal system and the methods according to which it was administered were without parallel in other parts of Madras. No survey or settlement was ever carried out in any part of the district. By 1945 some of the villages had come under the royotwari system. Prior to this, in 1917, the Agency Tracts Interest and Land Transfer Act was enacted which prohibited transfer of immovable property situated within the Agency tracts by a member of hill tribe unless made in favour of another member of a hill tribe, or with the previous consent in writing of the Agent or any other prescribed officer. Even though this law reduced the volume of alienation to some extent, unauthorized transfers were still prevalent and frequent. Thus, for quite sometime royati rights in land of tribals had been subject to certain restrictions. The realization of authorities that money-lenders and people mostly from plains easily induced the tribal tenants to part with their land with extremely inadequate considerations for transactions which were mostly fraudulent, resulted in the passing of another legislation in the post-independence era in Orissa. This legislation known as the Orissa Scheduled Areas Transfer of Immovable Property (by Scheduled Tribes) Regulation, 1956 commonly known as Regulation 2 of 1956 came as a subsequent effort to protect the interests of tribals. Under this regulation, STs cannot transfer their holdings to anyone who is not a member of any of the STs, without the permission of competent authority. Most of the tribals being illiterate and ignorant of their rights did not keep any written record of these transfers. Therefore, most of the transfers were verbal. Even though the Indian Registration Act, 1908 came into force during 1955-65 in phases in Koraput district, it was difficult to adduce any evidence for transfer. Thus, in many cases transfers had managed to mutate their names in records although such records were illegal.

With the above historical background, the present study was confined to two districts namely Koraput and Rayagada which are heavily and predominantly inhabited by tribals. The study was launched in three selected blocks in each of the two districts on pilot bais. Some of the gaps and deficiencies noticed in the existing revenue legal framework in general and those in specific in the study areas are analysed in the chapter

"Executive Summary and Recommendations". During the course of implementation of this project, a lot of challenges had to be confronted in terms of co-ordiantion, sharing, participation and involvement of all stakeholders. Even though the study areas are inhabited mostly by tribals where level of education has to go a long way in accelerating the pace of development in terms of education, health and infrastructure, the field staff engaged in the study exhibited tremendous courage and sincerity in collecting required information. The research team deserve all credit for this. I am grateful to the district authorities and the officers of connected departments who rendered unconditional and profound co-operation in sharing required data/information for the purpose of the study. The local NGOs/CBOs and individuals deserve all credits for their support.

We are extremely grateful to the Irish Embassy for their financial support in executing the programme in the districts. We are indeed thankful to the Dy. Chief of Mission of Irish Embassy and the functionaries for their unconditional technical and advisory support in making this project a reality.

I take this opportunity to place on record my deep appreciation for the help and assistance rendered by the research team. Our special thanks are due to the local institutions namely COATS, EKTA Koraput, NCDS, SCSTRTI, VASUNDHARA, EKTA PARISHAD at Bhubaneswar and OPDSC Rayagada for their help both in sharing information, data (secondary) and the office of Sub-Collectors Koraput and Rayagada, Revenue, Forest, Land Acquisition, R.I. Offices and the office of OSD, 1956 Regulation (2) of Koraput and Rayagada districts. The village committees, village leaders, SHGs and PRI elected members extended their unconditional support in every share of activities. I must express my profound gratitude to Gupteswar Panigrahy, advocate Koraput, Dr. P.C. Mohapatra, Dr. J.K. Samal of COATS Koraput, Durga Prasad Mohanty and Jagannath Mishra of EKTA Koraput for their sustained support during the project implementation in Koraput. I.S.S. is immensely obliged to B.S. Nayak, R.N. Satapathy and B. Panda advocates of Rayagada, R.K. Patnaik and N. Behera, Retd. Revenue Inspector for their sustained help as resource persons. My project colleagues Dr. B.K. Sahu, Dr.

N.K. Mishra, Dr. P.P. Patnaik, Manoj Kumar Swain, Sudhansu Mandal and Sanjaya Kumar Ghadai and the bunch of other village level facilitators and the trained tribal youths deserve special mention for their contribution to the making of this study. Last but not the least, I place on record the spontaneous and positive response of both the Collectors namely Gadadhar Parida of Koraput district and Dr. Nitin Bhanudas Jawale of Rayagada district in extending their help and physical presence on the occasion of distribution of *patta* (Record of Rights) to the people of different villages.

I hope and trust that this publication on this very special and key subject will be of immense benefit to those who are involved on the working of the subject matter representing academic and research world and those engaged in the land administration in the state and elsewhere.

Karunakar Patnaik
Senior Fellow & Regional Co-ordinator
Institute of Social Sciences, Bhubaneswar

Contents

List of Tables

List of Figures

List of Maps

Abbreviations

ADWO	-	Assistant District Welfare Officer
ATILT	-	Agency Tracts Interests and Land Transfer Act
CBOs	-	Community Based Organisations
CPLRA	-	Central Province Land Revenue Acts
DWO	-	District Welfare Officer
FGD	-	Focus Group Discussion
FRA	-	Forest Rights Act
GP	-	Gram Panchayat
GS	-	Gram Sabha
HAL	-	Hindustan Aeronautics Limited
HHs	-	Households
ISED	-	Institute of Socio-Economic Development
ISS	-	Institute of Social Sciences
LAA	-	Land Acquisition Act
MFP	-	Minor Forest Produce
MOU	-	Memorandum of Understanding
NALCO	-	National Aluminium Company
NGOs	-	Non-Governmental Organizations
NIRD	-	National Institute of Rural Development
No.	-	Number
NREGA	-	National Rural Employment Guarantee
NTFP	-	Non-Timber Forest Product
NTPC	-	National Thermal Power Corporation
OBC	-	Other Backward Classes
OEA	-	Orissa Estates Abolition Act
OGLS	-	Orissa Govt. Land Settlement Act
OLR	-	Orissa Land Reforms Act

OPLE	-	Orissa Prevention of Land Encroachment Act
OSATIP	-	Orissa Scheduled Areas Transfer of Immovable Property
OTA	-	Orissa Tenancy Act
OTRA	-	Orissa Tenant Relief Act
PAL	-	Panchayat at Appropriate Level
PESA	-	Provisions of the Panchayats Extension to the Scheduled Areas
PRIs	-	Panchayati Raj Institutions
PTGs	-	Primitive Tribal Groups
RCDC	-	Regional Centre for Development Cooperation
RI	-	Revenue Inspector
RoR	-	Record of Right
SC	-	Scheduled Caste
SCSTRTI	-	Scheduled Caste & Scheduled Tribe Research and Training Institute
SHGs	-	Self Help Groups
SPREAD	-	Society for Promoting Rural Education and Development
ST	-	Scheduled Tribe
UAIL	-	Utkala Alumina India Limited
UKB	-	Upper Kolab Project
UNDP	-	United Nations Development Programme
WEO	-	Welfare Extension Officer

Glossary

Adivasis: Tribals are otherwise known as *adivasis.*

Agriculture: Includes the raising of crps, grass or garden produce, horticulture, dairy farming, beeding and keeping of live-stock and use of land as pasture or for forest or for any other purpose where such use is ancillary to agriculture.

Alienation: Transfer of land from one to another in one or more forms.

Atala land: Plain areas of hill slopes, streams and nallas-adjoining areas.

Basti: Homestead land.

Benami: Dummy/fraudulent transactions.

Dalits: Scheduled caste people generally known as *dalits.*

Dangara: Lands on hill slopes, hillocks, stoney and hard type land.

Displacement: Displacement due to acquisition/transfer of land either or from homestead or agriculture land.

Estate: Includes a part of an estate and means any land held by or vested in an intermediary and included under one entry in any revenue roll or any of the general registers of revenue-paying lands and revenue-free lands, prepared and mentioned under the law relating to land revenue for the time being in force or under any rule, order, custom or usage having the force of law and includes revenue-free lands not entered on any register or revenue roll and all classes of tenures or under-tenures and any *jagir, inam* or *maufi* or other similar grant.

Fallow Land: Land suitable for cultivation.

Fig.: Figure.

Garden: Bagayat or orchard.

Gharbari: Homestead Land.

Gochar: Reserve land not normally allottable for lease or *patta* purpose.

Gram Panchayat: It is the lowest structure of the Panchayati Raj System consisting of Sarpanch, Panchayat Secretary and ward members.

Gram Sabha: It is the assembly of all voters of the Gram Panchayat.

Gram Sasan: The executive body of Gram Sabha.

Hac.: Hectares.

Holding: A parcel or parcels of land forming subject of a separate tenancy.

Homestead: By land, whether or not record as such, ordinary used as house-side, ancillary or incident to agriculture.

Jagir: Land given for free enjoyment in lieu of service rendered.

Khata: Document containing details of land owned by a tenant.

Landlord: A person immediately under whom land is held by a *raiyat* or a tenant.

Land Owner: A *raiyat*, a service tenure holder, a tenant, a landholder.

Land-holder: All holders or owners of interest in land between the *raiyat* and the state and a Proprietor, Sub-proprietor, Malguzar, Thikadar, Gountia, Tenure-holder, under-tenure-holder and includes an Inamdar, Jagirdar, Zamindar, Illaquadar Kharposhdar, Parganadar, Sarbarakar and Maufidar.

Landlessness: A stage of assetless-ness, ownerless of any type of land.

Lease: In relation to mines and minerals shall include a sublease a prospecting lease and agreement to lease and sublet and 'lease' shall be construed accordingly.

Paddy land: Soft, fully or partially irrigated, rainfed land.

Panchayat Samiti: It is the intermediary structure at the block level consisting of President, Vice President etc.

Patta: Local language of record of rights.

Public Purpose: Includes any purpose in relation to any common need, convenience or benefit of the village

Raiyat: A person who is deemed to be a *raiyat* as such (under the provisions of Orissa Land Reforms Act).

Revenue Officer: Any officer appointed as such by government to discharge any of the functions of a Revenue Officer (under the provisions of Orissa Land Reforms Act).

Royati: Tenanted land holdings.

Ryots: A person holding lands with rights of occupancy and deemed as such.

Standard Acre: The unit of measurement of land equivalent to one acre of class I land, one and one-half acres of class II land, three acres of class III land or four and one-half acres of class IV land.

Tenant: A person who holds land under another person, and is or but for a special contract would be, liable to pay rent for that land to that person.

Tenure: The interest of a tenure-holder or an under tenure-holder.

Thana: Police station.

Village: The area defined, surveyed and recorded as a separate village.

Zamindars: One type of intermediary.

Zilla Parishad: It exists in the district level consisting of President, Vice President etc.

Executive Summary and Recommendations

Land is the most precious asset for the lives of poor farmers in rural India. It is the basis for one's identity and symbol of social status within a family and community. The value of land for the well-being of people is ingrained deep in their consciousness. Therefore, no development activity can succeed unless primacy and priority is given to their persistent demand to recover lost lands and provide security of tenurial rights for the existing and prospective lands. Despite legal protection, land alienation has deprived many *adivasis*, *dalits* and other marginalized people of their entitlement to land due to displacement, mining, industrial and development projects, indebtedness, mortgaging of land, etc. leaving many in precarious conditions due to unrecognised leasing and share-cropping arrangements. Land creates inherent value for the livelihood by providing physical, social and nutritional security. The major issue closely linked to poverty in Orissa is the poor access and control over land among backward social groups. The distribution of land is highly unfair and unequal in the state where the state owns more land, while a large chunk of people remains deprived of access to land. The exclusion of rights and justice for getting rights over the land has been seen as the inheritance of historical, political and land governance system. Land rights can create the basis of access to other poverty-alleviation assets/livelihood outcomes. The real spirit of the land related laws especially tribal land alienation laws, are missing. Existence of parallel laws/Acts stand as major stumbling blocks in realization of rights of the tribals over land and livelihood resources.

Laws have been ineffective and inadequate. Through legal manipulation and other fraudulent means large-scale transfer of tribal lands to non-tribal people has taken place. In such a crucial juncture, it is apt that adequate legislations be made for protecting and preventing the tribal land alienation/ displacement/transfer for the improvement of their only source of sustainable livelihood. No doubt, a number of studies have been undertaken on this aspect, but the present study would sought find some of the gaps supposed to have been overlooked in other studies. Keeping the above in view, the present study was undertaken with objectives to (i) review and examine various land laws and regulations, (ii) to assess the present status of immovable properties, (iii) to identify the reasons of land alienation and to study the methods adopted for such alienation in the scheduled areas, (iv) to study the social, economic and psychological problem faced by the tribal women due to land alienation, (v) to study the role of Gram Sabha in preventing land, (vi) to assess the impact of land related laws especially relating to alienation/transfer/ displacement and (vii) to locate the gaps in implementation of these Acts.

The study was carried out in two scheduled districts namely Koraput and Rayagada of Orissa. Three blocks from each of the two districts were selected based on (i) the incidence of land alienation and displacement in and around the areas, (ii) consultation of key stakeholders. Keeping this in view, Koraput, Semiliguda and Laxmipur blocks from Koraput district and Rayagada, Kolnara and Bisamcuttack blocks from Rayagada district were selected for the purpose. From each block, two Panchayats were covered for the study. The selection of Gram Panchayats was done on the basis of suggestions given by the concerned officials, NGOs working on land issues and other key-stakeholders. Baseline survey was conducted to identify the issue-based villages. Two villages from each selected Panchayats were selected keeping in mind that the villages may have had a high incidence of land alienation cases based on base line survey report. All the physically existing households of the sample villages were covered in the study. Data was collected from both primary and secondary sources. Primary data includes (i) household survey, (ii) FGDs, (iii) semi-structured/informal interviews with the key stake-holders like elected Panchayat representatives, officials, activists, NGOs/CBOs, advocates and cross-sections of the society, (iv) case studies, (v) field observations and (vi) reality check/assessment. Secondary data were collected from revenue department, forest department records of state government, NGOs associated in land issues, population and agriculture data, district statistical handbook, study reports, published and unpublished records/documents etc. The present study suffers on account of the short duration of study period and due to illiteracy of respondents many of whom could not give required information properly.

The subject contents of this study report have been organised in seven chapters. Chapter-I is the introductory chapter covering the land related Acts/laws including prevention of land alienation Acts with special reference to land alienation in the tribal areas of Orissa starting from the history of land revenue system to the present land related Acts in force. The Chapter-II gives a picture on the context of the subject, objectives and methodology and area coverage of the study. Third chapter contains a brief review of different related studies. The fourth chapter focuses on the profile of the study area. Fifth chapter provides a broad layout of land holding and land use pattern in the study area. Land alienation and restoration explaining methods and causes of land alienation are mentioned in the sixth chapter. Seventh chapter critically reviews the impact of Acts/laws, gaps and suggestions. The overall study context/background, objectives, methodologies, area coverage and key findings with gaps and suggestions are reflected in 'Executive Summary and Recommendations' in the initial part of the book.

KEY FINDINGS OF THE STUDY

- In Koraput district, 2,439 households covering a population of 10,133 comprising of 48.9 per cent including female population of 5,178 (51.1%) were surveyed for the purpose. In case of Rayagada district, a total of 916 households having 4,120 population comprising of 1,923 (46.63%) male and 2,197 (53.33%) females were covered under the study.

- The *average size of family* is almost 4 in both the districts. Both the *age of the head* of household and size of the family was found to be low because sons got separated from their parents immediately after their marriage.
- In Koraput district, 91.43 per cent people are *Hindus* whereas in Rayagada district, except 5 households, all are Hindus.
- In Koraput district, *ST* (49.36%) concentration is more than the other castes like *SC* (13.57%), *OBC* (8.89%) and *general category* (28.17%). In Rayagada district, 87.12 per cent households are STs, 8.95 per cent SCs, 1.64 per cent OBCs and 2.29 per centhouseholds belong to general caste.
- Scheduled Tribe dominance is more in Rayagada as compared to Koraput district.
- In Koraput district, 29 per cent are literates. In Rayagada district the literacy rate is 38 per cent.
- While comparing *the literacy status* of both the districts, a significant gap is noticed between the two districts. This might be due to literacy calculation method. In Koraput district children between 0-6 years of age are included whereas in case of Rayagada district, the children between 0-6 years are excluded because of non-availability of authentic data.
- The study reveals that *female literacy* rate (29.29%) in the study villages of Raygada district is more in comparison to Koraput district which is 21.82 per cent.
- The study also revealed that literacy status is better in Rayagada district than in Koraput district.
- The rate of *workforce* is found to be 53.86 per cent in Koraput district and 42.23 per cent in Rayagada district. As workforce (both male and female) of majority of persons in Koraput district are reportedly engaged as wage-labour whereas more are engaged in agriculture and agriculture labour in Rayagada district compared to other occupations.
- While comparing the occupational pattern of *women*

workforce in both the districts, it was noticed that comparatively more number of women are working as daily wage-labour than other works.

- Majority of the HHs of the villages of Koraput district i.e. 50.4 per cent stand in the per annum *income range* of Rs. 10,000 Rs. 20,000 whereas in Rayagada district 39.52 per cent households are in the same income range. On the whole, between the two districts, the economic position in Rayagada district seems to be better than Koraput district.
- The *average expenditure* of HHs of study villages of Koraput district, on food i.e. Rs. 12, 690 annually which is more than other items of expenditure. The HHs incur very less amount on house maintenance, which is only Rs. 729 per annum. In case of Rayagada district, major portion of income is spent on food i.e. Rs. 12,880 annually and lowest amount i.e. Rs. 548 which is incurred on intoxicants.
- The study has revealed that in relation to the *nature of ownership* over land of households of Koraput district, 46.45 per cent of HHs were found to be landless. They were having neither *patta* land nor land transferred through inheritance. In Rayagada district, about 28.93 per cent are landless. On a comparison between the two districts, it is seen that land holdings are more in Rayagada district than Koraput district.
- In Koraput district, a total of 862 households possessed *Dangara* lands covering 3355.98 acres with an average holding size of 3.89 acres In Rayagada district, 321 households were having 2.12 acres of average holding size of *Dangara* lands with a total area of 681.36 acres. The average size of holding of *Dangara* land is found to be more in case of Koraput district than Rayagada district.
- Similarly, 575 households of Koraput district possessed *Paddy type land* covering a total of 826.34 acres with an average holding size of 1.44 acres. In case of Rayagada district, 204 households were having average size of 3.53 acres covering a total of 720.42 acres of paddy type land.

- In respect of *Atala* land, 40 households of Koraput district possessed *Atala type land* with an average size of 1.70 acres covering a total of 68.18 acres. Similarly, in Rayagada district, 240 households possessed 738.12 acres of *Atala* type of land with average size of 3.07 acres.
- In Koraput district, out of the total sample households, 337 households occupied 483.48 acres of *encroached land.* Among the encroachers, 283 households occupied 432.24 acres of *Dangara* land, 52 households 49.52 acres of Paddy land and 2 households were having 1.72 acres of *Atala* land. In Rayagada district, 321 persons were encroaching an area of 601.45 acres of land out of which 223 persons were occupying 403.39 acres of *Dangara* land, 21 persons encroached 36.06 acres of Paddy type land and 77 persons encroached 162 acres of *Atala* type land.
- While analyzing the position of *lease land* in Koraput district, it is seen that out of total 31 lease land holders covering 75.29 acres of land, 17 persons received *Dangara* type of land covering 29.13 acres and 13 persons got paddy type lands covering 45.16 acres. Only one person has one acre of *Atala* land. Similarly, in Rayagada district, out of 52 total lease *patta* holders covering 75.55 acres, 7 households received 7.75 acres of *Dangara* land, 21 households 40 acres of Paddy land and 24 households received 27.80 acres of *Atala* land.
- With regard to the *average size of land holding* and extent of area covered in the study villages of Koraput district, it was found that the average size of land holding is 4.92 acres with a total area of 4259.49 acres. In case of Rayagada district, the average size of land holding is 4.08 acres covering total of 2139.9 acres of land.
- In *operationalisation of land holding* in Koraput district, 59 per cent are found to be marginal farmers, 18 per cent small landholders and 22 per cent large landholders. In case of Rayagada district, 54.68 per cent of the households are small farmers, 26.39 per cent marginal farmers and 18.93 per cent households large

farmers. Between the two districts, marginal farmers are found to be more in Koraput district than Rayagada district.

- Among all categories of *transfer of land* in Koraput district, maximum transfer was found to have been effected through mortgage (78 Nos.) followed by sale (55 Nos.) and 82 through other methods. In respect of Scheduled Tribe households in particular, the same trend was noticed i.e. maximum through mortgage (51 Nos.) followed by sale (21 Nos.) and 49 numbers through other methods.

  In the study villages of Rayagada district, it revealed that altogether 78 cases of transfer had taken place, out of which 41 households transferred land through mortgage, 22 households through sale, and rest 15 households through other methods of transfers like oral transfer, encroachment/forcible occupation, public purpose, land acquisition, cheating and *benami* transfers etc. In both the districts, transfer was effected mainly through mortgage.
- In Koraput district, out of total 55 *land alienation cases through sale*, 13 HHs reported that they alienated land for purchasing bullocks, 12 households alienated land to meet household expenses, 10 HHs to meet medical treatment and rest 20 HHs alienated for other purpose. In Rayagada district, out of 22 land alienation cases through sale, seventeen households for household expenses while in respect of 2 households, alienation was done for repayment of debt. In the remaining cases, alienation was due to social and religious reasons, medical treatment and inability to cultivate land.
- The *reason for mortgaging* of land in the study villages of Koraput district in respect of 21 HHs out of total 78 HHs, was to meet requirement for medical treatment, 15 households for purchase of bullock, 11 HHs for meeting household expenses and the rest for other reasons. In Rayagada district, out of 41 transfers through mortgage, 20 households reported to have mortgaged to meet expenses for agricultural

improvement, 8 to meet household expenditure, 5 for debt repayment, 3 to meet expenses in social and religious functions and the rest for other purposes.

- In Koraput district, out of total 391 cases of *restoration of lands*, 284 cases were instituted on application filed and 107 by *suo motu* initiation by competent authorities. Out of these, in 182 cases, lands were restored back during the period 2005-08. In Rayagada district, out of total 568 cases (including regulation 2 of 1956 and OLR, 1960) in 238 cases, lands were restored back during the period 2005-08.

Gaps and Recommendations

- Orissa Scheduled Areas Transfer of Immovable Property (OSATIP) 1956, Regulation (2) followed by 2002 amendment further restricts transfer of tribal land. The objectives of these regulations are very good for ensuring better livelihood sustainability of the tribals. At the same time, the regulations prevent poor tribals from selling their land even during emergent situations arising out of financial difficulties.
- The Act is not effective in cases where transactions take place without any written evidence, which is otherwise called verbal transfer.
- Lack of awareness about the regulation creates problems in its effective implementation. Hence, awareness about different land related laws especially laws preventing transfer of tribal lands should be generated among the community.
- The role of Gram Sabha in land alienation under PESA is not effective.
- Many Sarapanches are not aware of PESA where role of Gram Sabha is very crucial.
- During cross verification in Koraput district, it came to notice that in some cases, lands have been restored back as per government records, but the original land owners have not been informed about such restoration and therefore, are not in actual physical possession of these lands.

- Government have distributed *patta* in respect of lease lands but the majority of *patta* holders are not in possession of those lands and they have not been shown the land physically by demarcation or physical possession of land has been effected.
- In Rayagada district, some of the transferees are still in possession of lands. When the tribal victims were asked about their interest in getting back lands, they abruptly replied that parties have already constructed houses on the land transfer and staying thereon at present. Since they have no other houses to stay, it would be an injustice to claim for return of those lands.
- The Regulation (2) of 1956 read with amendment in 2002 should be made more effective through proper and regular monitoring during its implementation.
- Further amendment to the Regulation is needed in the interest of the tribals to meet emergent situations.
- It is very difficult on the part of government to update the mutation cases because the number of legal heirs have increased even more than 50 in a family in some cases. As a result, mutation cases cannot be settled even at government level. So survey and settlement operation should be initiated immediately for updating record of rights.
- Competent revenue authorities seem to be not paying special attention to cases of restoration of illegally transferred lands of tribals to non-tribals. *Suo mottu* initiation is restricted and limited as revenue authorities show little and scant attention to such cases. The production of proof and evidence to establish such claims may be further liberalized to enable competent authorities for better and expeditious disposal of such cases.
- For restoration cases, baseline enumeration, village-wise may be undertaken and recorded at RI office level. In the regular revenue meetings, the progress of restoration cases may be reviewed court-wise and on pending cases, date lines and targets be fixed for disposal.

- Alienation of land in different forms could be assessed at the tehsil level and position reviewed.
- Allotment of land to landless in respect of agricultural and homestead land should be channelised in a systematic manner with involvement of PRI institutions.
- Land records may be updated through e-network at the RI level.
- Encroachment cases could be categorized after baseline survey and classified as settleable with reference to existing land position in villages and disposed of in camp courts within specific time frame.
- On the whole, land laws and issues are many and complicated. The need for a comprehensive, self-contained and consolidated revenue code with substantive and procedural provisions applicable to all categories of people including ST and SC is essential now.

1

Introduction

This chapter is the introductory chapter containing a brief historical background of land administration and review of some land related Acts and laws with special reference to land alienation in the tribal areas of Orissa starting from the history of land revenue system to the present day land related legislations in force.

BACKGROUND

Land is the source of sustainable livelihood and food for almost half the world's population. It is one of the primary necessities of humans. It not only gives a sense of belonging but also provides a source of livelihood. Understanding the real value of land, people always try to possess as much land as possible. Although land and land-based resources are more vital to the livelihoods of tribal people, tribals have, in reality, poor access to land and forests. Forest dwellers, tribals, marginal farmers and rural population are very much dependent on land for agriculture and other related livelihood options. Furthermore, land, forest and other local natural resources are not only the livelihood resources but also are generally equated with social, economic and political power. The landowners and encroachers of land often belong to the upper class of the society. The cultivators normally belong to the lower middle caste/class and agriculture laboures mostly the tribals and weaker sections that own no or little land. They work as agricultural labourers for survival without any security. The women are the worst affected. They don't have a say in decision-making relating to control and management of land resources.

By the time of independence, India inherited a large tribal population. The tribals who are the original inhabitants or first settlers of India and who have contributed much to its culture, history and heritage, have become foreigners in their own native land and made victims of development, displacement and human-rights violations. Tribals are normally the people living in groups in a closer society in or near the forests far remote from the plains. Tribal economy is traditionally centred on land and other land-based resources with exception of small communities within them who live as hunters and food gatherers even today. More than 90 per cent of the tribal people depend upon agriculture and allied activities. As cultivators and agricultural labourers they work on land and as 'adivasis' they collect food and minor forest produces to earn their living. They have been leading an almost anonymous life in forest although they were enjoying rights and privileges for utilizing land in the forests they lived in. To them in rural and forest areas, land is the basic asset and they attach much importance to land not because it is their main stay but for propitiation of their several social practices and rituals. Land is, therefore, organically linked to the lives of the tribals and it is a precious resource for them. Among most of the tribal communities, individual or family rights are subsumed under community rights. It means that the village community has supremacy over land resources and rights of individual families are secondary. The social status of the tribal is determined to a considerable extent on his right over the land particularly the hereditary land. This traditional basis of ownership of land continued until the beginning of the British rule. From ancient times, right to ownership of land was conferred on customary basis to those who cleared dense forests with strenuous efforts. In order to integrate the tribals into mainstream of the society, a lot of efforts have been taken in the past. But all these have resulted, on the contrary, in their cultural alienation, dehumanization and deprivation of livelihood resources even though tribal culture, identity and ideology, tribal development and tribal rights are the issues that form the core of discourse on the tribals of India. (Gnana Prakasham, 2004). Regrettably, the tribals are at the bottom of social and political ladder in India. Development projects have not only by-passed them, but have often harmed them by taking away their lands and other resources on which their livelihood was based. Therefore, it is important to understand that tribal communities are vulnerable because of their poverty, asset-less-ness and illiteracy compared to the general population.

Prior to the intervention of the outside agencies, the tribal people had free access to the land owned by their respective communities. Individual ownership of land was unknown in their customary rule. Right to use land in conformity with the decisions of community leader was in practice. The unplanned loss of natural resources due to imbalances created on land and land-based resources around their natural environment have become a severe threat to tribal subsistence. Rudimentary technologies and techniques of production innovated, developed and adopted by the tribal people in the past appeared to be less productive in the perception of the non-tribal people but these continued to keep the socio-economic harmony in tact within the tribal community. Even though there are several problems relating to lands particularly of the tribals, the question is not just how much land these people have lost at present and how much land has been alienated in to the hands of non-tribal households, and/or to the State in the name of development projects. The way tribal people have lost their land and have been deprived of this, is a serious issue before the social anthropologists and in particular the development workers in general who are engaged in tribal development. In Indian society, more particularly among the tribal communities, loss of land seems to be the loss of social identity not only for the present generation but also for the future generations.

The survival of tribal communities has a vital stake on land and forest resources. For historical reasons, most of the tribal people inhabit the forest and hilly regions of Orissa. The social structure and day-to-day living of tribal people reflect their intimate relationship with land and forest surrounding their habitat. In the context of Orissa, it has been observed that different tribal communities are inhabited in specific locations. The inflow of non-tribal population started in large scale into the tribal areas when the British started their contact with these regions. This process also facilitated inter-territorial movement between neighbouring tribes for acquiring natural resources for livelihood. In this process, the original settlers in a particular territory asserted their rights over all the surrounding natural resources, and the neighbouring tribes hardly questioned this. However, subjugation over tribal groups and control over their resources by colonial government and native kings (Rajas) at different points of time has been questioned, and the agony of the tribal people has been reflected through resistance and rebellions, but the tribal people were always made subdued and subservient by the non-tribals.

It is believed by some that the tribal communities in India originally lived in the fertile belts of the Indo-Gangetic plain and river valleys. When the Indo-Aryan speaking people penetrated into the country, they had a series of encounters with these tribal communities; and by means of their superior technology and military power, pushed the tribals into hills and forests and settled down themselves in the fertile belts. The tribal communities who were cultivators, distributed lands among themselves according to individual needs and cultivated allotted land on a cooperative basis. There was no system of measuring land for the sake of revenue collection, and there was no authority to which any share of crop was paid as tribute, tax or revenue. In course of time, this was replaced by individual land tenure under the overlordship of feudal chief because of the limited dimension of the kingdoms and direct contact of the chief with the peasantry. The process of feudalisation of the older agrarian structure based on chieftainship was set in motion under the Mughal empire. Having little knowledge about the country and being constantly engaged in warfare, the Muslim rulers introduced an intermediary system in land revenue administration to collect land revenue. The Hindu kings and chiefs formed the group of intermediary landholders and received grants from Muslim and later Mughal ruiers in the form of *iqtas* for the revenue and military services they rendered to the state.

The feudal revenue system heavily taxed the people. In

order to maintain their magnificent court, large army and pompous style of living, the Muslim rulers invariably needed a large amount of money. They made big demands for land revenue from land proprietors who in turn, resorted to rack-renting from cultivators in order to meet the demands of their sovereign. When Muslim rule weakened in the 15th century, the system of beneficial tenure became entrenched, and the hereditary *iqtdars* were free to lease out tax collection, which intensified rack-renting and made the economic life of the people even more unbearable. Finding the oppression unbearable, many small landowners had to abandon their holdings in favour of *zamindars* and became their dependents and agricultural servants. This is the beginning of the process by which people particularly tribals lost land due to the *zamindars* and other categories of intermediaries. Another type of alienation during British regime took place, when they started infrastructure projects like construction of railways and roadways for transportation of raw materials to the port cities of India and further transhipping them to England. They acquired land both from the tribals and the non-tribals, at their sweet will. They also acquired land for establishment of military cantonments. Thereafter, the non-tribals who entered into tribal territory during the middle of the nineteenth century led the tribals into debt-trap. During British regime the tribals of Bihar, Orissa and other parts of India revolted against government establishment and the outsiders (Dikus) for illegal acquisition of their land. Kohl insurrection of 1832, Santal Hull (Rebellion) of 1855 and 1877, Birsa movement of 1895 are the examples of tribal unrests against the British and the Hindu landlords. Baba Tika Majhi who fought against the British during later part of eighteenth century is not accorded the status of freedom fighter. All the mentioned unrests took place prior to Sepoy Mutiny, 1857. Non-tribal historians (Hindu historians) have called these tribal movements as 'peasant movements'. Baba Tika Majhi, who shot dead Augustus Cleveland, the then Deputy Commissioner of Bhagalpur was an imaginary person in the opinion of Hindu historians.

After these movements, British regime passed some Acts in the Imperial Parliament to restrict transfer or alienation of the tribal land to non-tribals. Most important among these

Acts are Chotanagpur Tenures, Act, 1869, Chotanagpur Tenancy Act, 1855 and Santal Parganas Tenancy, Act, 1875. In order to regulate the transfer of tenure land of the tenants for "public purpose' a law was passed in British Parliament which is known as Land Acquisition Act, 1894. This Act is still in operation for easy alienation of tribal land for public purpose. Alienation of tribal land is continuing at the same pace since British regime, despite plethora of positive land legislations in favour of tribal people (Majhi, U.N, 2009).

The Government of India had made a number of plans from time to time to protect the rights and security of the tribals and to integrate them into national developmental planning. Tribal people's rights, being a part of the broad human rights phenomenon, have acquired significance in recent times. As human beings the people who live in tribal areas deserve to enjoy a similar kind of rights like others. Being citizens of the country, they are also entitled to get several benefits. But in due course of time, tribal villages and settlements have disappeared and the tribals are deprived and dispossessed of their livelihood resources. They are displaced, the communities are fragmented and decisions are taken without their involvement. Projects are planned and policies are formulated without their knowledge. All this has resulted in their further impoverishment, marginalisation and human-rights violation. In contemporary society, the states come to their rescue to a limited extent.

HISTORY OF LAND REVENUE SYSTEMS IN ORISSA

The method of land revenue system implemented by Todar Mal during Akbar's regime incorporated measurement, classification and fixation of rent as its main components. Under the various pre-British regimes, land revenues collected by the state confirmed the rights to land produce, and that it was the sole owner of the land. British rulers took a nod from this system and allowed the existence of non-cultivating intermediaries. The existence of these parasitic intermediaries served as an economic instrument to extract high revenues (Dutt, 1947) as well as sustaining the political hold on the country.

The different parts of Orissa, historically, have inherited different land revenue administration systems from Bengal Province (northern Orissa), Madras Presidency (southern Orissa), Central Province (western Orissa), and the former (24) princely states. Large areas of the total area of the State fell under *zamindari* systems, in which many layers of 'intermediaries' between the landlord and cultivator were responsible for exacting land revenue. *Ryotwari* (peasant-proprietary) systems prevailed over parts of southern Orissa that had been under the Madras Presidency.

REVENUE ADMINISTRATION UNDER BRITISH ORISSA: 1905-1947

In British Orissa, there was no comprehensive revenue law for the province. Only a list of practices or rights jotted down in the course of a settlement report by the settlement officer, or a document of conditions imposed on the headman of a village, extending to an enumeration of the rights and liabilities of *raiyats* were limited to explain the legal position of revenue law. In the early days, the cultivators were a few and the cultivation was very little. The right of land was not so much insisted upon and the cultivator was satisfied with merely the permission to enjoy the land. As more land was available than people to cultivate, the cultivator did not think of retaining an ownership, although it was of no help but to drag him to the system. All that the cultivator wanted was to enjoy the fruit of his labour, and liked it to be wherever he could spend his labour on a piece of land.

As population increased, the demand for cultivation of more land also increased. As a result, the greediness or tyranny of the *zamindars* or rulers also increased with the fear of losing the land. While the strength of the headman/*zamindar* increased, the strength of the cultivators declined. Loss of land became a serious matter for the cultivator and with it, the need for acquisition of right over the land.

Under *zamindari* (or landlord) tenure, land was held as an independent property and revenue was assessed on an individual or a community owning an estate as a landlord. Proprietors were required to deposit land revenue at the district treasury. One sub-divisional officer, assisted by tehsildars was in-charge of revenue collection. There was no revenue administration below the district level, and the *zamindars* organised their own revenue collection agencies, often involving many more layers of intermediaries. There were two main systems of land revenue assessment in Orissa prior to independence: *Zamindari* and *Ryotwari*. The former existed in five districts in Orissa, the latter in a part of a district, and the so-called subsidiary alliance in the 24 princely states covering the remaining seven districts. More than 80 per cent of privately owned land fell under the *zamindari* system (Pathy, 1981). Under the *Ryotwari* (or peasant proprietary) tenure, land belonged to the Crown and was held in a right of occupancy (which was both heritable and transferable) by individuals. Revenue was assessed on individuals who were the actual occupants of smaller holdings. It was collected through the village headman whose office was hereditary. He was paid a commission (10 per cent) and sometimes received some *jagir* (service) lands. In addition to collection of land revenue, he was also required to keep the records-of-rights up-to-date by carrying out mutations. There were numerous rent-paying sub-tenants under either of the system.

A number of different revenue and tenancy laws were also prevalent in the state. This was because parts of Orissa fell within different administrative units belonging to Bengal (later with Bihar after its separation from Bengal in 1912), Central Provinces, and Madras. The British, through a subsidiary alliance by which the princes had freedom in their internal administration so long as they paid regular tributes to the colonial authority, controlled the 24 princely states in Orissa.

LAND TENURE AND REVENUE SYSTEMS IN ORISSA AT INDEPENDENCE

Bengal revenue system: It covered northern part of the state, comprising the undivided districts of Cuttack, Puri, and Baleswar (but excluding the princely states subsequently merged in these districts). In these areas, the Bengal Rent Act, 1859 was the first legislative attempt to regulate tenancy, replaced by Bengal Tenancy Act, 1885. In 1913, the Orissa

Tenancy Act was modelled more or less on the Bengal Tenancy Act. Many intermediary forms of tenure subsequently developed in these *zamindari* areas, and an increase in share-cropping was suggested which remained in vogue from that period.

Madras revenue system: It extended over the southern part of the state, comprising the undivided districts of Ganjam, Koraput, and Baliguda sub-division of Boudh-Khondmal (now Kandhamal) district i.e. Oriya-speaking areas of the Madras Presidency. Here the first attempt on tenancy legislation was the Madras Estates Land Act, 1908, which applied to the *zamindari* areas of Madras Presidency. There were also *ryotwari* areas under the state government where the rights of landholders were governed not by law but by executive instructions contained in the Board's Standing Orders, which had the force of law. As in *zamindari* areas, landholders (*ryots*) could freely sub-let to tenants who had no protection under the law.

Central province system: This system prevailed across the western part of state, comprising the undivided districts of Sambalpur and Nawapada i.e. the Oriya-speaking areas of former Central Provinces. In these areas, the Central Province Land Revenue Act, 1881 and 1917 and the Central Province Tenancy Act, 1898 and 1920 governed land revenue and tenancy systems.

Princely states: These partially excluded areas had separate land settlement/revenue regulations under the Government of India Act, 1935. There were no written laws designed to protect the interests of tenants in most of the princely states. The Orissa States Order, 1948 conferred occupancy rights on tenants, but no rights were recognized for any tenants below occupancy tenant in the hierarchy of rights in land.

Debottar grants: The number of temple grants particularly in Orissa state was enormous on which the state lost thousands of rupees of land revenue on account of them. The *debottar* grants were controlled by the rulers personally or by members of the ruling family. Usually, there was no audit of these expenditures. The misappropriation of *debottar* funds in Athagarh by the rulers is mentioned in the report of that state

(Ex-state). The *debottar* funds constituted an addition to the income of the ruler or ruling family. Any attempt to interfere with the administration of these grants will, of course, result in the cry of interference with religion. The grants to deities or temples, or *debottar* grants as they were called in Orissa, are also extremely extensive in the then Orissa province.

Bramhottar grants: These are grants to the Brahmins. Most of these are held rent-free and the *sanads* give what may be construed as absolute proprietary rights, though in practice, transfer is not permitted in most ex-states or is restricted to Brahmins. Rulers in former days encouraged immigration of these men of learning and religion, considering it necessary not only for the well-being and culture of the country-side, but also for their own spiritual good.

In view of the divergent revenue laws prevalent in the state, the question arises as to what, if any, was the impact of different revenue systems prevalent in the state until half a century ago? While there may be little practical difference, rights over commons, which are determined by traditional norms and customs, these do vary across the state, and especially between the erstwhile *ryotwari* and *zamindari* areas. There was also considerable variation in the quality of land records management since there was a village accountant in *ryotwari* areas, but no such position in *zamindari* areas. As a result, land records were better maintained in the former and almost non-existent in the later. These differences were likely to create problems during land litigation since there may be no historical records on which to establish the basis of competing claims. The settlement or recovery of ceiling lands was very difficult. *Gaontias* or petty *Zamindars* were controlling large chunks of best quality of land. It is still prevailing in some parts of the western and southern Orissa.

LAND TENURE AND LEGISLATIVE INITIATIVES IN POST-INDEPENDENCE ORISSA

During centuries of exploitation and dominance by the feudals and arbitrary nature of company provincial laws, there were realisations and revolts for change in economic policy and social structure of the state for providing equal and secured rights for adequate means of livelihood to its citizen. As far as land rights were concerned, it was a fight against the feudal structure and concentration of wealth in a few hands. The continuation of the intermediaries had completely ceased the access of land to the tiller and actual cultivators. After independence, several initiatives were taken up and new legislations enacted. The basic objectives of these new legislations was to reduce poverty and hunger by protecting the interests of the marginalized with equitable distribution of resources. The land reforms initiative was one of the major steps taken to fulfil the set goals.

LAND REFORM INITIATIVES

The basic principles of land reforms initiative considered by Planning Commission in the 'First Five Year Plan' was abolition of intermediary interests so as to bring the *ryots* into direct relationship with the state. Keeping the aforesaid objectives in view, some of the legislations framed in this direction are analysed below:

The Orissa Estates Abolition Act, 1952

The harmful effects of the continuance of intermediary interest like *malgujar, thikadar, gauntia, inamdar, jagirdar, sabarakar* and *mafidar* was seriously felt. In many parts of Orissa, the owners did not cultivate the land themselves. A large amount of damage was done to agriculture due to the fact that *zamindars* (intermediaries) were giving land to under-tenants or *bhagchasis* and sub-letting of land, frequently evicting tenants from cultivated lands and absence of the concept of land to the tiller. Keeping in view the situation, government had appointed a 'Land Revenue and Land Tenure Committee' under the Chairmanship of N.K. Choudhury in 1946, which recommended an integrated scheme of land reforms in the first phase for elimination of intermediaries. The outcome of the committee's recommendation was the Orissa Estates Abolition (OEA) Act, 1952. The Act provided for abolition of all rights, title and interest in land of intermediaries by whatever name known, including mortgages, lessees of such interest between the *raiyat* and the state of Orissa, for vesting in the said state of the said rights, title and interest. However, the validity of the vesting order of the OEA Act was challenged by the *Zamindars* of Ganjam, Nagra (Sundargarh), Hemagiri and Sarafgarh, Madhupur, Kanika for which much progress in the abolition of the estates could not be made up to 1954.

The main provisions of the Act are: (i) Abolition of intermediaries, (ii) Vesting of all land rights in the state and allowing agricultural land of less than 33 acres to remain with the intermediary for personal cultivation. Mearns and Sinha, while reviewing existing land legislations found that the Act aimed at abolishing intermediaries but did not contain any provision for protecting the tenants. Large-scale eviction of tenants took place as *Zamindar* was allowed to resume land less than 33 acres for personal cultivation. Owing to the absence of reliable records, abolition of intermediaries could not be completed until 1974.

Orissa Tenant's Relief Act, 1955

This Act was a temporary measure for protecting tenants including sub-tenants of crop-sharers from ejectment and fixing maximum rents. The tenants were protected from ejectment subject to the landowner's right to resume land for personal cultivation. If the landlord, after evicting a tenant for personal cultivation, failed to personally cultivate the land during the first quarters of the next agricultural year, the tenant was entitled to restoration. Orissa is one of a few states in India that had attempted legally to abolish tenancy (land-leasing), except in the case of persons of disability the definition of which includes widows, divorcees, and other unmarried women. As a result, land rights may pass on to any cultivator who can demonstrate continuous occupation over a period of at least 12 years (adverse possession). While tenancy remains widespread, these restrictions led to concealed forms (e.g. oral contracts), which give tenants little or no protection in law. A ceiling on individual land holdings also applies, and currently stands at 10 'standard acres' (depending on land quality). In addition to these provisions, which fall under land reforms legislation, three major Acts govern land administration, and respectively provide the basis for land survey and settlement, land consolidation/prevention of land fragmentation, and prevention of encroachment on government land. These are briefly described below.

ORISSA LAND REFORMS (OLR) ACT, 1960

The land reforms programme of the first and second five year plans had two objectives: (i) to remove such impediments to agricultural production as arose from the tenurial structure inherited from the past and (ii) to eliminate all elements of exploitation and social injustice within the agrarian system to provide security to the tiller of the soil and assure equality of status and opportunity to all sections of rural population.

The First Five Year Plan did not set any guideline in the matter of upper ceiling limit and a land-owner was to be permitted to hold for his personal cultivation. Fixation of ceiling limit was deferred to the subsequent plans for want of detailed information on agrarian structure, distribution of land and relationship between different sections of rural population.

The OLR Act, 1960 was a landmark legislation enacted to envisage better rights to weaker sections of the tenancy and fixation of ceiling on holdings of the land-owners. It was a progressive legislation relating to agrarian reforms and land tenures after the abolition of intermediary interest. The main objectives of the Act were:

- Introduction of uniformity in the land rights or *raiyats*.
- Provision of better rights in favour of temporary lessees, share-croppers and tenants.
- Conferment of occupancy right in homestead lands.
- Settlement of disputes between landlord and tenants.
- Regulation of rent.
- Protection of Scheduled Caste and Scheduled Tribe *ryots* from illegal alienation of land.
- Resumption of land for personal cultivation by landlords.
- Determination of non-resumable land areas of tenants and fixation of ceiling on land holdings.

Main provisions of the Act are:

- Permanent, heritable and transferable rights in land for the tiller.
- Ban on leasing of land except under special conditions.
- Adverse possession of land in continuous cultivations for 12 years or more by a person other than its owner to pass on to the cultivator.
- Rent not to exceed one-fourth of the gross-produce.
- Ceiling on individual holdings at 33 standard acres – (later reduced to 20 in 1965) and to 10 standard acres in 1972.

The impact of the law according to Mearns and Sinha are as follows:

- Delay in the enactment and actual implementation of the Act provided sufficient opportunities for large landholders to escape ceiling restrictions.
- By explicitly banning tenancy, the law swept the problem of share-cropping under the carpet. No provision was made to record concealed tenancies.

ORISSA GOVERNMENT LAND SETTLEMENT ACT, 1962

In 1961 and then in 1966, lease principles were declared which laid-down the guidelines for settling government land. Priority was given to landless scheduled tribes, scheduled castes and other backward castes. Settlement of land with tribals could be done up to an extent of 5 acres. These provisions were formalised through the Orissa Government Land Settlement Act, 1962 (applicable from 1965). This Act was later amended in 1974 [section 3(e)] which allowed the revenue officials to settle cultivable wastelands to cultivators subject to the restriction that 70 per cent of such land in a village was to be settled on priority basis with scheduled tribes and scheduled castes [section 3(2)]. Such settlement was to be made in the following priority areas namely cooperative farming societies of land-less agricultural labourers, any other landless labour, ex-serviceman or members of armed forces who are residents in the village and any farmer who does not cultivate more than one standard acre of land [section 3(3)]. However, uncultivable wastelands (abada ajogya anabadi) were treated as not fit for agriculture and were, therefore, termed as "objectionable" for the purposes of settling them for agriculture (Govt. of orissa instructions for reservation of government land for specific purposes in rural areas dated 28 Jan.1966). Thus, uncultivable wastelands were not to be given out for agriculture purposes but could be settled for other purposes.

ORISSA PREVENTION OF LAND ENCROACHMENT ACT, 1972 (AMENDED IN 1982)

Another important legislation, which has bearing on government lands, is the Orissa Prevention of Land Encroachment Act, 1972, which governs eviction or regularization of encroachments on government land.

The largest extent of government land is held by the revenue department under the following four categories:

- *Abad Jogya Anabadi* (*cultivable waste lands*): *Abad Jogya Anabadi* (cultivable waste lands) include cultivable land within village boundaries not settled with *ryots*, abandoned holdings, surrendered holdings, land reserved for Panchayats and armed forces personnel. These lands are "unobjectionable" for regularization and can be settled with raiyats.
- *Abad Ajogya Anabadi* (*uncultivable waste lands*): It includes water bodies (including rivers, lakes and sea), hillocks, mountains, hilly areas and sandy areas. This category often forms the most important category of non-forest government land in tribal areas where hills and sloping areas have been categorized as *abad ajogya anabadi*. This category is also treated as "objectionable" for settlement as it is presumed to be uncultivable.
- *Rakshita* (*reserved lands*): It includes lands acquired under Land Acquisition Act, 1984 but not transferred, irrigation works, lands transferred to gram panchayats, *gochar* lands (grazing land), land for village settlement, *parambok* land, *gramya* jungle etc.

- *Sarvasadharana* (*communal land*): It is defined in section 2 (b) of the Orissa Government Land Settlement Rules, 1983, as communal land used by village for community purpose such as graveyards or cremation ground, places where festivals or *melas* are held, playgrounds, temples, mosques and other religious places of worship, village roads and other land reserved for public purposes.

The major provisions of the Act are: (i) unauthorized occupation of government land prohibited, (ii) penalties on encroachment to be followed by eviction, (iii) settlement of two (later amended to one in 1982) standard acres of "unobjectionable" land (i.e. govt. wasteland) with "eligible" beneficiaries (e.g. landless).

In the case of encroachments on revenue wastelands, only landless persons are eligible for allotment of 'unobjectionable' land for agriculture purpose [Section 7 (a.b)]. Under this Act, landless encroachers, though legally entitled to claim rights on non-objectionable lands, must have first committed an unlawful act i.e. encroachment and the Revenue Inspector should detect the case and report it to the tehsildar under section 3(1) of the Act and the Orissa Prevention of Land Encroachment Rules, 1985. The tehsildar is empowered to initiate the proceedings in the encroachment case, ascertain if the encroacher is to be evicted or is eligible for land settlement by virtue of landlessness and then settle the land in the name of the encroacher (Section 15, OPLE Rules, 1985). Only then, and after due process of law, can the landless be issued with the *patta*, which regularizes the land rights. The entire process hinges upon whether or not the Revenue Inspector reports the encroachment in the first place and whether the Tehasildar initiates the case proceedings. There are a few or no incentives for these officials to do so and therefore frequently bribes are demanded or eviction threatened (Mearns and Sinha, 1998).

The central purpose of the Orissa Prevention of Land Encroachment Act, 1972 is to prevent unauthorized occupation of government land. It is even-handed and clearly lays down penalties for all instances of encroachment, to be followed by eviction. Through a 1982 amendment, the state has recognized that there are different types of encroachers on different types of lands. While all encroachers are required to be evicted from "objectionable" lands, landless encroachers are legally entitled to be settled on up to one standard acre per family of "non-objectionable" land (i.e. cultivable wasteland).

The eviction process proceeds as per the following procedure (i) Based on his enquiries, the RI files a report to the land revenue court (tehsildars) and legal proceedings are initiated, (ii) The Tahsildar issues notice to the encroacher to show cause why penalty should not be imposed, (iii) Encroacher shows such cause, (iv) If this cause is admitted as valid, the settlement process is initiated for the issue of *patta* of the land, (v) If cause not admitted as valid, notice is issued for imposition of penalty and eviction within 30 days.

The statutory minimum period for the entire process is three months, but in practice cases may take between six months and a year or even more to settle/dispose of. Steps (1) to (3) in the above sequence are similar for both landless and land-owning encroachers on "unobjectionable" land. Only landless encroachers may move to step (4) whereas land owning or other ineligible encroachers are imposed penalty and eviction notices issued as in step (5). As described earlier, however, the incentives for RIs and other tehsil-level officials to 'book' encroachment cases in this manner are very low as encroachment represents a lucrative source for rent seeking. Therefore, actual cases reported are less than the actual encroachment, taking place on the ground.

Besides allotting land to the landless and other eligible categories (under Orissa Government Land Settlement Rules, 1983) for agriculture purpose, the government can put this land to use for development projects, or allot to industries, projects, government departments, state or private corporations, local statutory bodies, or cooperative societies (Schedule-II, Orissa Government Land Settlement Act). In doing so, the government is not accountable to inhabitants of the village in whose boundary these lands are situated, and who may have been using the land for agriculture or common purposes. This has been a serious problem with acquisition and transfer of land, including government land for development projects or forests without consent of the local communities

LAWS FOR PREVENTION OF TRIBAL LAND ALIENATION

Alienation of land in tribal areas is one of the important causes of poverty among tribals. Alienation may refer to (i) displacement in the name of development and conservation projects, (ii) sale and transfer of land by tribals to other tribals and non-tribals, (iii) alienation of land due to indebtedness, (iv) forcible eviction of tribals from their land and unauthorised occupation of tribal land by non-tribals or by public authorities, (v) conversion of land from communal ownership to individual ownership, (vi) increasing urbanization, (vii) treating tribals traditionally occupying forest land as encroachers, (viii) government land allotted to tribals under various schemes but substantive possession not given, (ix) developments taking place close to the tribal habitats which force the tribals to move out though there is no formal transfer and acquisition of land.

In order to prevent alienation of tribal lands to non-tribals; various laws were passed in different parts of pre-independent Orissa. Most of the princely states regulated transfer of land owned by tribals to non-tribals. In areas directly ruled by the British with tribal populations, there were laws and regulations to check tribal land alienation. In the agency areas under Madras Presidency, the Agency Tracts Interests and Land Transfer Act, 1917 was applicable to protect the interests of *adivasis* in land. It provided that all transfer of land by a member of hill tribe to a non-tribal without permission of competent authorities was null and void. But most of the laws were ineffective, full of loopholes and implemented only in default (Mishra and Behera, 2005). For example, the list of communities to be considered as hill tribes in ATILT, 1917, included Malis and Gaudas (currently classified as other castes) as hill tribes between 1927 and 1950. Even Dombas (currently classified as scheduled caste) were included as hill tribes during the period 1927-1943. This meant that during this period, any transfer from tribal communities like Kondha or Gadaba to Gadabas/Malis or Dombs was valid and did not attract the provisions of this Act.

PREVENTION OF TRANSFER AND RESTORATION OF STs/SCs LAND UNDER OLR ACT

Under section 22 of OLR Act, 1960, any transfer of holding by a *raiyat* belonging to scheduled tribe shall be void only except when it is made in favour of a person belonging to STs or SCs to SCs and if transferred by STs/SCs to non-STs/SCs without permission of the competent authority or by the revenue officer in areas other than scheduled areas. Under sections 23 and 23A of OLR Act, the revenue officer or competent authority shall act *suo motu* or on the application of interested person for restoration of the property to the transferor or his heir, if it has been transferred without prior written permission of the competent authority or if the land has been under un-authorised occupation by non-SC and non-ST persons. If one goes to the *in situ* experience of detection and restoration of alienated land in rural areas and more in backward tribal areas, the process of *suo motu* initiative by revenue officer or any competent authority for that matter is almost non-existent. The figures of illegal land transfer found in government statistics are extremely meagre to the ground situation. The revenue department, instead of being a solution setter becomes a part to the problems. In fact, restoration of property means actual delivery of possession of the property to the transferor or his heir. But there are cases of conflicts relating to the possession of property restored to the transferor or his heir. The restoration is made in official papers but the property is actually not given possession of in the field, depending upon various factors of the local power structure or other structural dynamics. Hence, the objectives of land reforms initiative get vanquished.

TRANSFER OF LAND

The Orissa Scheduled Areas Transfer of Immovable Property (by Scheduled Tribes) (OSATIP) Regulation (2) of 1956 is the principal legislative measure which regulates the transfer of immovable property of schedule tribes in schedule areas of the state. This is popularly known as Regulation (2). This regulation was promulgated by Governor of Orissa with immediate effect in exercise of powers conferred by sub-paragraph 2 of paragraph 5 of the 5th Schedule of the Constitution of India. The regulation received the assent of President of India on 21st September 1956 and was 1st published in Orissa Gazette (Extraordinary) dated 4.10.1956. (*Detail in Appendix-I*)

Regulation 2 of 1956 has clearly adopted the definition of scheduled areas made in the Scheduled Areas (part-A - States Order, 1950 and scheduled tribes lists specified in respect of the state of Orissa in the constitution of India (Scheduled Tribes) Order, 1950. It defines the meaning of transfer of immovable property and competent authority who can give permission to transfer of land from the hands of tribal people to non-tribal people. It also explains the eviction of persons from unauthorized occupation of property, inadmissibility of deed for the purpose of registration, surrender or relinquishment not to be valid unless settled by the landlord, bar to attachment of immovable property, various offenses, refund of consideration money in case of invalid transfer, burden of proof and power of court to ignore administration, amendment of the Limitation Act, 1963 in its application to the scheduled areas, and the amendment so far taken up by the state in the regulation 2 of 1956 in respect of procedure for transfer to non-scheduled tribe, attachment of immovable property, control of district collector, sale of timbers by scheduled tribes and scheduled castes standing on their homestead (gharbari)/*royati* land.

The Regulation 2 of 1956 was amended in December 19, 1957 in order to have effective implementation and to safeguard the interests of the tribals. The important provisions of the amended regulation include, extension of the period of limitation from 12 years to 30 years, relaxation on the transfer of land by way mortgage for obtaining loans for agricultural purposes and also in favour of co-operative societies for the said purpose, emphasis on the restoration provision of rigorous imprisonment for a term which may extend to two years or with fine which may extend to Rs. 2000 or both, provision of higher penalty for summary eviction, and has further been provided that the consideration money paid by the transferee to the transferer for such invalid transfer shall not be refunded. Further, the amendment also states that steps are being taken to safeguard the interest of scheduled tribe in the non-scheduled areas of the state by making necessary provision in the Orissa Land Reforms Act of 1960 in the light of amendments to the Orissa Regulation 2 of 1956.

REGULATION 2 OF 1956 VIS-À-VIS ORISSA LAND REFORMS ACT, 1960

Because of faulty implementation of OSATIP Regulation and the OLR Act, 1960, there was need for amendement of sections 22 & 23 of Orissa Land Reforms Act, 1960 in order to prevent indiscriminate transfer of tribal land. It was decided by government that the competent authority should invariably submit the case record to the concerned collector and obtain his approval before passing final order for granting permission. This should be mandatory and statutory. Hence, suitable provisions may be incorporated in sections 22 & 23 of O.L.R. Act, 1960 and in Regulation 2 of 1956. The section 22 of O.L.R. Act, 1960 should prescribe specific conditions for granting permission for sale and/or transfer of tribal land. Further, in order to streamline granting of permission, a definite or fixed time limit should be prescribed. Section 22 of O.L.R. Act, 1960 should provide a minimum ceiling beyond which no permission is to be granted for transfer of tribal land. Transfer of tribal land, transfer of immovable property in favour of a female member of a scheduled tribe who is married to a person who does not belong to any scheduled tribe should be deemed to be a transfer made in favour of a person not belonging to a scheduled tribe and this provision may be incorporated in sections 22 and 23 of the O.L.R. Act, which is analogous to Regulation 2 of 1956.

Both the enactments, however, are silent about the case where transfer made in favour of a male member of a scheduled tribe who is married to non-tribal woman shall be deemed to be made to a tribal. This requires incorporation in the O.L.R. Act and Regulation 2 of 1956 in the shape of suitable amendments (Behura and Mohanti, 2009). Although Orissa Scheduled Area Transfer of Immovable Property (OSATIP), Regulation 2 of 1956 has been in force it has come to the notice that large-scale alienation of tribal land to non-tribals has been made in the scheduled areas of the state and as such it has become a matter of great concern for the state government as well as the Government of India. Keeping this in view, the said Regulation 2 of 1956 has been further amended to read as the Orissa Scheduled Areas Transfer of Immovable Property (by Scheduled Tribes) Amendment Regulation of 2000, which was assented to by the President on the 20th August, 2002 and was published in Orissa Gazette in form of notification on the 4th September, 2002. The Regulation was made by the Governor of Orissa under sub-paragraph (2) of paragraph 5 of the Fifth Schedule to the Constitution of India. The amendment emphasizes on the fact that such transfer of immovable properly shall take place among members of STs and not in favour of any non-ST person. Further, the total extent of land for such transfer was enhanced from minimum one acre to two acres in case of irrigated land and five acres for non-irrigated land. A non-tribal man married to a tribal woman shall not be eligible for transferring land under this clause. The amendment contains various executive instructions for effective implementation of the regulation. As per the amended provisions of the said regulation, transfer/alienation of land of STs to persons not belonging to STs has been completely banned. Any such transfer shall be null and void if the same has been made without written permission of the competent authority. In case any transfer has been made in contravention to the provisions in the regulation, the competent authority either *suo motu* or on a petition filed on that behalf, shall declare such transfer as illegal and shall restore the land to the lawful landowner or his/her heirs following the prescribed procedure. The regulation also provides for eviction of persons in forcible occupation of land belonging to members of STs and restoration thereof. The regulation provides for penal action in respect of illegal transfer as well as unauthorised occupation.

This amended provision restricts the basic principles of constitution as well as scheduled area administration. However, the existing law (Regulation 2) to check illegal transfer of land in scheduled areas in Orissa seems to be weaker than similar laws enforced in other states like Andhra Pradesh where no land in scheduled areas can be transferred to non-tribals, whether private or government land (Study Report, 2007). The Orissa Scheduled Area Transfer of Immovable Property (OSATIP) Regulation 2 of 1956 merely restricts transfer of *patta* land of tribals to non-tribals. However, looking at the size of land holding in scheduled areas, 84 per cent of the land area of scheduled districts either belongs to the government or to non-tribals. Large areas of land customarily claimed by scheduled tribes were categorized as forest land or revenue land through survey and settlements and forest declarations. The tribals privately own only 16 per cent of the land. There have been huge instances (unreported) of illegal transfer of tribal land to non-tribals in violation of the protective regulations (OSATIP 1956) on which the government is supposed to give special attention as its special constitutional responsibility. In a sample verification of illegal transfer it was seen that as per the officially reported cases of illegal transfer only 318 cases (out of 434) of land restoration were disposed of under the existing regulation during 1st April, 1999 to 31st March, 2000 (one year) in Kandhamal district. This shows that if the existing OSATIP regulation is sincerely implemented it can give enough protection to the tribals (Study Report, 2007). The government was under the process of amending that provision of the regulation for sale and transfer of tribal land to non-tribals or for other development purposes keeping in view the necessity of transferring the land to the multinational mining and industrial houses.

RESTORATION OF ALIENATED LANDS

The Orissa Scheduled Areas Transfer of Immovable Property (by Scheduled tribes) Regulation, 1956 and Orissa Land Reforms Act, 1960 provide for restoration of illegally alienated land to the tribals. Sections 23 and 23(A) of Orissa Land Reforms Act, 1960 provide for restoration of the land to the tribal owners. In case of land transfer in contravention of section 22 of OLR Act, the revenue officer on his own information or on the application of any interested person may declare such a transaction invalid and restore such property to the owner or to his heirs. Section 23 (A) of the OLR Act provides that if a non-tribal person is found to be in unauthorised occupation of land of a tribal person by way of trespass or transfer in contravention of laws, the revenue officer may order eviction and shall cause restoration of such property to the tribal owner or his heirs. Sections 3(2), 3(A), and 7(A) of the Regulation 2 of 1956 also provide for restoration of any land transferred against the law. Section 3(2) empowers the competent authority to order ejectment of the non-tribal person and restore the land to the tribal owner or his heirs. In case, restoration is not reasonably practicable, the authority may settle the property with another member of the scheduled tribe or in the absence of any such member, with any other in accordance with the provisions contained in the Orissa Government Land Settlement Act, 1962. Section 7(A) of the Act states that despite an eviction order, if a non-tribal person is found to be in re-occupation of the immovable property of a tribal person, the former shall be punishable with rigorous imprisonment or fine imposed by the appropriate court of law.

THE PANCHAYATS EXTENSION TO SCHEDULED AREAS (PESA), ACT, 1996

To provide self-rule for tribals, a Bill was introduced in Parliament in December 1996 based on the recommendations of a parliamentary committee called the Bhuria Committee after a thorough examination of the tribal scenario in the country. The Bill was passed by the Parliament and became an Act receiving the President's assent on 24th December, 1996. The Act is known as the Provisions of the Panchayats (Extension to the Scheduled Areas) Act, 1996 (No. 40 of 1996) popularly known as PESA Act (*See Appendix-II*). This is not an amendment of the constitution. It has been made clear in Article 243(M)(4)(b) that extending the provisions of Part-IX of the Constitution shall not be deemed as an amendment to the constitution under Article 368.

The objectives of the Act are: (i) to extend the provisions of part IX of the Constitution relating to the Panchayats in the scheduled areas with certain modifications, (ii) to provide self-rule for the bulk of the tribal population, (iii) to have village governance with participatory democracy and to make the

Gram Sabha a nucleus of all activities, (iv) to evolve a suitable administrative framework consistent with traditional practices, (v) to safeguard and to preserve the traditions and customs of tribal communities, (vi) to empower Panchayats at the appropriate levels with specific powers conducive to tribal requirements, and (vii) to prevent Panchayats at the higher level from assuming the powers and authority of Panchayats at the lower level of the Gram Sabha. The Act is applicable to the scheduled areas as referred to in clause (1) of Article 244 of the Constitution. Under the extension Act of 1996, the Gram Sabha has been the nucleus of all activities and endowed with specific and wide-ranging powers and functions, which do not figure in the main provisions of part-IX of the constitution. At the same time, the Gram Sabha in scheduled areas is also required to manage natural resources including land, water, forests within the area of the village in accordance with provisions of the constitution and other relevant laws for the time being in force. With the passing of the Panchayats (Extension to Schedule Areas) (PESA) Act, 1996 (Act No. 40 of 1996), the state government of Orissa amended the Orissa Gram Panchayat Act, 1964, Orissa Panchayat Samiti Act, 1959 and Orissa Zilla Parishad Act, 1991 w.e.f. 22.12.1997, i.e. within the mandatory period of one year of coming into force of PESA Act effective from 24th December, 1996. In the amended law, the following subjects have been placed within the purview of Panchayats at respective tiers.

GRAM PANCHAYATS

The Gram Panchayats have been empowered to exercise such powers and perform such functions to the extent as may be prescribed in respect of the following matters (i) enforcement of prohibition or regulation or restriction of sale and consumption of any intoxicant, (ii) ownership of minor forest produce, (iii) prevention of alienation of land and restoration of any unlawfully alienated land of a scheduled tribe, (iv) control over money lending to scheduled tribe, (v) subject to control and supervision of Gram Sasan (Gram Sabha is the executive authority of Gram Sasan) Gram Panchayat shall have the power to manage village markets.

PANCHAYAT SAMITI

Similarly, the Panchayat Samiti, in consultation with the Gram Sasan, shall be competent (i) to exercise control and supervision of the nature and extent as may be prescribed over institutions and functionaries of various social sectors in relation to programmes and measures as may be prescribed by the government, (ii) to prepare local plan including tribal sub-plan for the area and to exercise control over resources of such plans.

ZILLA PARISHAD

The Zilla Parishad has been empowered to see that (i) no prospecting license or mining lease for minor minerals by auction shall be granted under any law on or after the amendment except with prior recommendation of the Parishad, (ii) no acquisition of land for development projects or for resettling or rehabilitation of persons affected by such projects shall be made under any law without prior consultation with the Parishad.

LAND ACQUISITION

The Land Acquisition Act, 1894 read with 1984 empowers the government to acquire any land for 'public purpose', with adequate compensation wherever necessary. The central PESA Act mandates that there should be consultation with the Panchayat before land acquisition for development projects and before resettling or rehabilitating persons affected by such projects. The Orissa government acquired large areas of tribal lands in the past for industrial, mining and other purposes, but except in a few cases, the displaced people were left in sway owing to payment of inadequate/delayed compensation and/or inadequate rehabilitation measures. The displacement of people from their land actually causes displacement from their livelihood because some of them loose best quality agricultural lands while some others are affected by loss of forests rich in minor forest products. While some people are deprived of facilities for marketing their products others have no scope to work as agricultural labourers.

The uprooting of the tribal people from their homes and habitat for building dams and other industrial development projects has been one of the shocking scandals of post-independent India. It is a fact that sufficient attention has not been given to this problem. Providing monetary compensation has not been of much use. Not only was it inadequate, but also the tribal people were given lump sum amounts of money, which they could not use properly. They were left with nothing in a short period of time. Even to make claims for alternative land they had no records of their ownership or titles in many cases. The rehabilitation projects were flawed as the tribal people were put in areas, which had no similarity with the habitat they were used to live in. They were given often rocky or barren land. Displacement has meant that the evacuated tribal people are driven to take up back-breaking jobs as construction labour working in brick kilns and other forms of labour in the unorganized sector.

Even though the Indian constitution provides for protection of rights in land for scheduled tribes, in Orissa, the scheduled tribes have been affected by large-scale displacement due to dams, industrial and mining projects. Land for land as compensation was generally not provided, and displacements which have led to severe cultural, social and economic deprivation of tribals.

The PESA Act has made it mandatory to consult the Panchayat before granting prospective license on mining lease for minor minerals as well as granting of concession for exploitation of minor minerals by acquisition. Despite enactment of PESA Act, all matters concerning land acquisition for projects and leasing out of land for minerals are decided in accordance with the existing Acts mostly the Land Acquisition Act of 1894 as amended in 1984 and the Coal Bearing Areas (Acquisition and Development) Act, 1957. These Acts, do not assign any role or power to the Gram Sabha in granting lands and leases and do not attach any importance to rehabilitation and resettlement of affected person. However, recently the Central Government and the state government of Orissa have come out with their 'Rehabilitation and Resettlement Policies' in 2006 and 2007 respectively. But in reality, the provisions of PESA seem to have watered down further by operation of the "Eminent Domain" concept under the Land Acquisition Act of 1894.

The power of prevention of land alienation and restoration of illegally alienated land under the PESA have been vested with both the Gram Sabha and the Panchayat at appropriate level. One of the crucial reasons why a necessary mandate to the Gram Sabha or any tier of local self-government institution has been envisaged is the degree of significance the central government attaches to certain subjects that are critical to the lives of the tribals. Land alienation is one such critical aspect among others in the context of common property resources. Different states have responded differently. While the Gujarat government has only involved the district Panchayat, the Maharashtra government has mandated that it shall be competent for every *Gram Sabha* in the scheduled areas to make recommendations through *Panchayat* having regard to the provisions of any law for the time being in force pertaining to transfer or alienation of land of the persons belonging to the scheduled tribes. The government shall be competent to make suitable recommendations to the Collector in this regard. Here again the precedence has been given to the already existing laws in the state rather than the spirit of the new PESA Act. The laws enacted for protection of the tribal lands in the scheduled areas under the fifth schedule of the constitution by the states have not prevented alienation of tribal lands and large-scale transfers as such. Orissa Regulation 2 (1956) in scheduled areas provides that land held by a person belonging to a scheduled tribe cannot, without permission of the appropriate authorities, be alienated to a person not belonging to a scheduled tribe. Further, government of Orissa has amended the aforesaid regulation again by Regulation (1) of 2002 which has made the regulation more stringent to the extent that no tribal owner can transfer land to even another tribal if the total land remaining with the tribal is less than two acres of irrigated land or five acres of un-irrigated land. In spite of these provisions, due to the loopholes in the laws, interference of the bureaucracy and political authorities have subverted whatever legal safeguards exists in the Acts and regulations. Non-tribals continue to take over tribal lands through different methods such as mortgages, lease agreements, *benami* transfers, false title deeds in collusion with revenue officials, by marrying tribal women or holding/ recording such lands in the name of the bonded tribal agricultural labourers. To put an end to this problem, the government of Orissa has made necessary amendments to the Regulation 2 of 1956, which came into force with effect from 4.9.2002. As per the amended provisions of the said regulation, transfer/alienation of land belonging to STs to non-tribals has been completely banned. Any such transfer shall be null and void if the same has been made without written permission of the competent authority. In case any transfer has been made in contravention of the provisions in the regulation, the competent authority either *suo motu* or on a petition filed on that behalf, shall declare such transfer as illegal and shall restore the land to the lawful landowner or his/her heirs following the prescribed procedure. The regulation also provides for eviction of persons in forcible occupation of land belonging to members of STs and restoration thereof. The regulation provides for penal action in respect of illegal transfer as well as unauthorised occupation.

LAND ACQUISITION ACT, 1894 READ WITH 1984 AMENDMENT ACT

The Land Acquisition Act, 1894 read with Amendment Act of 1984 is the basis on which the government acquires private land. The Act also prescribes the procedure for acquiring property. This Act (1894) was inherited from the colonial rulers, who obviously viewed land in a very different context. The Act itself was based on the theory of 'eminent domain' that the state is sovereign and it has supremacy over all natural resources and is best placed to "protect" or exploit the same. Thus the state can compulsorily acquire any property including private by the process of law. This is a central Act and is applicable to all the states though the states are allowed to make changes (amendments) as long as such changes are not opposed to the provisions as they stand in the Act. There are regional variations with regard to matters such as who can set the process of acquisition in motion, manner of notices to be sent, persons on whom notices must be served etc. As per the

Constitution of India, the states having Schedule-V areas are allowed to make changes in the Act (or any other Act for that matter), through a public notification issued by the Governor of the state after consultation with the Tribes Advisory Council of the state and these regulations take effect after obtaining the consent of the President of India. In addition to the L.A. Act, 1894, the Coal Bearing Areas (Acquisition and Development) Act, 1957 also permits the Central Government to acquire land for coal mining purpose. The Coal Bearing Areas Act gives more power to the state and it is administered directly by the Central Government by appointing a competent authority. Most of the rules and procedures in this Act are similar to the newly enacted Land Acquisition Act of 1984.

The Act was amended only after 37 years of Independence in 1984. Other than bringing into its fold the states of Rajasthan, Kerala and Nagaland, which till then had self-contained Land Acquisition Acts, the 1984 amendment was made primarily with a view to streamlining the process of land acquisition. The important amendments made relating to the process of acquisition are (i) A time limit is set down for the completion of all formalities between the issue of preliminary notices under section 4(1) and the issues of the declaration of acquisition under section 6(1). This time limit is fixed as one year, (ii) Time limit within which the Collector must make an award as in section 11-A is fixed at two years from the date of publication of the declaration under section 6(1), (iii) Payment of compensation with interest at 12 per cent per annum, commencing from the date of notice under section 4(1) to the date of the Collector's award under section (II-A), (iv) Enhancement of the payment of solarium (i.e. compensation for loss, suffering or injuries) from 15 to 30 per cent of the market value of the acquired land, (v) Land acquisition for private companies has been bought into the ambit of this Act: This is probably the most drastic amendment, which allows the private companies to approach the government to acquire land for setting up their industries and related enterprises and services.

While some of the above changes can prove beneficial to the displaced as they minimise undue delay and also provide for compensation at a higher rate, another clause included under section 17 confers more discretionary powers on the Government for acquiring land in urgent cases. This clause that gives the Collector the powers and the right to acquire land after 15 days of publication of the notice under section 9(1). The amended Act also allows for treating acquisition for government company or government controlled corporations including 51 per cent (government controlled) joint ventures as public purpose. This obviously simplifies the process of acquisition for such companies and corporations. Typically the procedure for acquiring land for a company is the same as when acquiring land for a public purpose. However, no steps can be taken unless the company has obtained the previous consent of the government and the company has executed an agreement under section 39. This agreement would cover the following matters and should be published in the official gazette. These are (i) payment to the government the cost of acquisition, (ii) transfer of the land to the company on payment of such cost, (iii) terms under which the land is held by the company.

One of the most contentious issues in land acquisition has been the issue of "public purpose" and the state government, through its district administration has almost discretionary powers in this regard. The Act does not provide any exhaustive definition of "public purpose" and merely gives examples of the kind of things which can be considered as 'public purpose' including (i) provision of village sites, or the extension, planned development or improvement of existing village sites, (ii) provision of land for town or rural planning, (iii) provision of land for planned development of land from public funds in pursuance of any scheme or policy of government and subsequent disposal thereof in whole or in part by lease, assignments or outright sale with the object of securing further development as planned, (iv) the provision of land for a corporation owned or controlled by the state, (v) provision of land for residential purposes to the poor or landless or to persons residing in area affected by natural calamities, or to persons displaced or affected by reason or the implementation of any scheme undertaken by government, any local authority or corporation owned or controlled by the state, (vi) provision of land for carrying out any educational housing, health or slum clearance scheme sponsored by government or by any authority established by government for carrying out any such scheme, or, with the prior approval of the appropriate Government, by a local authority, or a society registered under the Societies Registration Act, 1860, or under any corresponding law for the time being in force in a state or a cooperative society within the meaning of any law relating to cooperative societies for the time being in force in the state, (vii) provision of land for any other scheme of development sponsored by government, or, with the prior approval of the appropriate government, by a local authority, and (viii) provision of any premises or buildings for locating a public office, but does not include acquisition of land companies.

PESA AND LAND ACQUISITION LEGISLATION

The provision in the PESA Act relating to land acquisition mandates for a consultation with the Gram Sabha prior to acquisition of land. Land Acquisition Act being a central Act, the onus is on the Central Government to make the relevant modifications. However, in 1998, the Ministry of Rural Development issued executive instructions to all the chief secretaries describing the modalities of consultation with the Gram Sabhas and detailing the procedure for land acquisition in Schedule-V areas. The summary of executive instructions issued on land acquisition in the light of PESA Act included the following issues (i) Gram Panchayat-wise schedule of land proposed to be acquired, (ii) A separate letter of consent from each of the Gram Panchayat in favour of the proposed acquisition. This should be in the form of a written resolution of the Gram Sabha, (iii) In case a Gram Sabha expresses a disagreement, a copy of the resolution showing the reasons for disagreement including alternative suggestions, if any, made by the Gram Sabha.

The district collector is required to examine the letter of consent and continues to hold the right to over-rule the dissent of the Gram Sabha, but is expected to follow procedure providing for a tripartite meeting with the villagers, government and interested party. In case, even after this issue is not resolved, the collector can proceed with the acquisition as provided under the Land Acquisition Act. The instruction also details the responsibilities of the state government for coordination and monitoring of land acquisition and rehabilitation and re-settlement schemes in the scheduled areas. These instructions require the government to maintain a register showing year-wise quantum of land acquired in the Schedule-V areas and the formation of an inter-ministerial coordination committee for monitoring. This committee may publish a district-wise lands acquired for public purpose. It is however; not very clear if these instructions have been meticulously followed by the concerned authorities. In most cases, Land Acquisition Act, 1894 (as amended in 1984) is the only legislation, which has been used for the acquisition of the land for development projects. This law provides for cash compensation for the acquired land and structure on it. The complex system of compensation envisaged under it denies the legitimate rights and interests of the claimants. The Act recognises only individuals. It does not recognise the collective or community rights and hereditary usufruct (the use and profit) rights. Again, the Act recognises only the records of rights. As administration have occupied government land for generations together but do not have any records of right or have traditionally depended on the Common Property Resource (CPR) or have survived by rendering services to the village community as a whole, are not entitled to any compensation under the present legislation.

LOSS OF LAND AND LIVELIHOOD DUE TO DISPLACEMENT: REHABILITATION FRAMEWORK

In the name of overall national development the national and state governments are in consistent move for providing land to the mining and industrial companies besides land for conservation projects. These lands are mainly acquired and given from the poverty-stricken backward tribal areas, which are basically resource rich, meaning thereby making the poor tribals more vulnerable and impoverished. Nearly 1.5 million people were displaced by development projects between 1951 and 1995, of which 42 per cent were tribals. As per this estimate, less than 25 per cent of the displaced tribals were ever resettled even partially. The marginalized population mostly tribals and *dalits* have been cultivating over generations for almost all of these lands, where no attempt was made to settle their rights through survey and settlement process or under Orissa Prevention of Land Encroachment Act (OPLEA), 1972.

Orissa is an extremely mineral rich state most of which lie in the tribal districts. The hilly terrain and availability of water in the area also makes them suitable for constructing reservoirs and dams. The major dams taken up in scheduled areas are the Machkund, Salandi, Balimela, Upper Kolab, Indrawati, Mandira etc. in Orissa. The major industrial projects like Rourkela Steel Plant, NALCO's Alumina refinery at Damanjodi, HAL, Sunabeda have been taken up. Large number of industrial projects are under implementation and the proposed projects are the Alumina Refineries of UAIL in Kashipur, Vedanta at Lanjigarh, Tata Steel at Kalinga Nagar, POSCO at Paradip, etc. However, the state government, of late, has formulated a clear rehabilitation policy namely "Orissa Resettlement and Rehabilitation Policy 2006". The objectives of the government in general are (i) to avoid displacement where possible and minimize it, exercising available options otherwise, (ii) to facilitate resettlement/ rehabilitation process, (iii) recognize voices of the displaced communities (emphasizing the needs of the indigenous communities and vulnerable sections), (iv) ensure environmental sustainability through participatory and transparent process, and (v) help guide the process of developing institutional mechanisms for implementation, monitoring, conflict resolution and grievance redressal (*See Appendix-III*). The lack of legal title of tribals on most of the lands cultivated and occupied by them for generations has aggravated situations leading to serious law and order problem where either payment of compensation or rehabilitation is required for establishment of title over acquired lands. The new Rehabilitation and Resettlement Policy of both the central and state governments require sincere implementation.

A cursory look to the policy indicates several pro-active and positive promises and if implemented sincerely could result in providing several livelihood options to the oustees, particularly the tribals.

FOREST RIGHTS ACTS, 2006

The enactment of the "Scheduled Tribes and Other Traditional (Recognition of Forest Rights) Act, 2006, in short, Forest Rights Act (FRA)" is a revolutionary step taken by the central government. Enactment of FRA 2006 is a watershed in the continuing hard-fought struggle of tribals and other forest dwellers of India. For the first time in the country, the state formally recognizes the historical injustice done to forest-dwelling communities during the consolidation of state forests. This new legislation (FRA) attempts to right the historic wrong and empowers the right holding communities and Gram Sabhas to "protect, conserve, generate and manage" the community forest resources for sustainable use. This legislation is meant to address both forest rights and occupancy rights of the people who had been using forest lands for generations without any formal recognition. The Act does not distribute lands to the people but merely recognizes the rights of forest dwellers over land that they have been residing in and cultivating. In this Act, the recognized rights of forest dwelling scheduled tribes and other traditional forest dwellers also include responsibilities for sustainable use, conservation of biodiversity and maintenance of ecological balance for strengthening the conservation regime of the forests while ensuring livelihood and forest security for the forest dwelling scheduled tribes and other traditional forest dwellers. Several rights are declared as forest rights of forest dwelling scheduled tribes and other traditional forest dwellers on all forest lands (*See Appendix-IV*).

The Scheduled Tribes and Traditional Forest Dwellers (Recognition of Forest Rights) Act, 2006 (Central Act) particularly the forest right committee under it creates an anomalous position in regard to the scheme of Joint Forest Committee Management (V.S.S.) of forest department. This anomalous position be removed. The village committees should be set free from the bonds of joint forest management to enable them exercise their legitimate rights over the new Central Act.

WOMEN AND LAND RIGHTS

Women's land rights have often been subsumed as those of her husband's. She is not counted as an owner in her own right and this makes her disproportionately vulnerable to losing her entitlement over land. As per 1988 prevention of *benami transaction* law, if there is land in the name of the wife or daughter, it would be assumed that the land belongs respectively to the husband or father. Significant gender bias persisted both in land reform legislation and in personal laws. Most of the land reform legislations were enacted during the 1980s before women's land rights were considered worthy of policy attention (Agarwal, 1994). No mention of women's land rights was made until the Sixth Five Year Plan (1980-85). The Eighth Five Year Plan (1990-95) called for a change in inheritance laws to accommodate women's rights but gave few specifics and called upon state governments to allot 40 per cent of ceiling-surplus or state-redistributed land in the name of women alone, with the remainder to be jointly titled.

Women and land rights is one of the most crucial issues of any society that is trying to bring in equity and justice. Access to land is a necessary condition for poverty alleviation. Therefore, an adequate legal framework for land rights to women, is more essential. In the context of a developing country, land rights are crucial as they are linked to issues like right to food, work and other human rights. It is now increasingly understood that the denial of inheritance of land rights especially in a patriarchal system has contributed to the subordinate status of women. Therefore, the Hindu Succession Act, 1956 as well as several other legislations have been amended keeping in mind a gender-just approach. The past has revealed that even when the government made efforts for the distribution of degraded and ceiling surplus land to both the husband and wife in the case of landless persons, this either was not implemented or hurdles in implementation were not shared. As a result, the beneficiaries would be unaware of the system of joint *patta* as the application for land settlement came normally in the name of the husband. Therefore, a fresh circular in 1989 made the joint application both in the name of husband

and wife mandatory failing which land *patta* would not be granted. In the case of a widow, the *patta* would be in her name.

In order to give preference to widows, unmarried women, victimized women and women living below the poverty line, in 2002, the government of Orissa decided to allot at least 40 per cent of the government wasteland kept for agriculture and house site purpose, ceiling surplus land and *bhoodan* land. The government also stressed that at least 40 per cent of this land should be allotted to women belonging to the scheduled tribes and scheduled castes. The recent Orissa Rehabilitation & Resettlement Policy, 2006 has some progressive features such as unmarried daughters/ sisters more than 30 years of age should be treated as separate families. Physically challenged persons, orphans who are minors, widows and women divorcees are also to be treated as separate families. Though elaborate provisions have been made to ensure land rights to women, in practice, this has not always been followed. There are significant gaps between women's land rights and their actual ownership and possession and between the limited ownership rights and their effective control over land (Bina Agarwal, 1994). For instance, the Orissa Land Reforms Act, does not mention the order of devolution of tenancy land-whether it would be according to personal law or would follow a different order. This is not clear and therefore open to interpretation. Women's legal rights in land conflict with deep-seated social norms and customs, and are rarely recognized socially to be legitimate. There are strong pressures, reinforced by social stigma, seclusion practices, and other sanctions, on women to forfeit their legal rights in favour of their brothers. Women tend to internalize the imposed social restrictions because they feel that they may find themselves dependent on their brothers for economic and social support in the event of widow-hood or marital break-up. Women's lack of control over independent sources of income has implications not just for their own well-being, but also for their children's, since it is known that child nutritional status is more closely related to women's than men's income. To the extent that women's income is land-based, women's lack of access to and effective control over land may, therefore, threaten the well-being of many household members. This applies to women from all social strata. On efficiency grounds, women are often the sole or *de facto* heads of households and, on the assumption that greater tenure/title security provides production incentives, granting them independent title to land is likely to lead to higher agricultural output.

However, this is only applicable if existing gender biases in agricultural support services and market factors are corrected. Indeed, recent experience with savings and loans groups in India suggests that women are frequently better credit risk managers than are men. In the context of women's role in agriculture, it would be worthwhile to mention about land rights provision for women in the Orissa Land Reforms Act. The Act allows widows, divorcees, unmarried women to lease their land for cultivation, when leasing is otherwise prohibited. Although this seems progressive, it may have two underlying presumptions (a) women should be protected from cultivation, therefore, they should be allowed to lease their lands, and (b) only female heads of households should have control over land, while the normal married women's control over land is subsumed under 'family'. Due to lack of education and awareness, women invariably do not have record of rights. One of the biggest limitations in matters of women getting the actual land rights is the lack of strict enforcement of law and strict monitoring of such empowerment process. For instance, the law says that if a family has more land than the ceiling set by the government, the surplus land is recorded in the name of the wife/daughter to avoid ceiling restrictions. Largely, it has been observed that people accumulate more than the ceiling and put it in the name of the women member of the family though *de facto* ownership lies with the male members. (Status Report, UNDP, 2008)

In the backdrop of historical evolution of land issues in India and in Orissa, there appears to be no point of hesitancy to admit that the tribals have had little or no access to land in spite of several legal provisions. Land as the means of livelihood had been adversely dealt with by the governance structure. Developmental activities withered away knowingly or unknowingly the rights and entitlements of tribals over land and in absence of sincerity, commitment and compassion to tribal causes, the tribals in the course of historical evolution, were deprived of their legitimate rights and entitlements over land and natural resources. The present study aimed at analysing the prevailing situation and place it open a few of the results that have emerged for redressal.

2

Objectives and Methodology of the Study

After a brief review of the existing land related laws in Orissa as well as in India, it has been observed that the real intention and objective of land-related legislations and provisions thereof are actually diluted, especially those intended to protect the tribals' livelihood and customary rights. The real spirit of the land related laws especially tribal land alienation laws, are missing. Existence of parallel laws/Acts stands as major stumbling blocks in realization of rights of the tribals over land and livelihood resources. Some of these examples are the Land Acquisition Act, 1989, the Forest Act and the Mining Act etc. Though each one has its own specificity on the question of land issue, they are very often clubbed together in relation to land acquisition and virtually tend to push the poor tribals to incomprehensible and irreparable loss in terms of livelihood options, habitat safety and economic parity. Similarly, the spirit of laws preventing transfer of tribal land under 1956 Regulation (2) has not properly attained its purpose.

RATIONALE OF THE STUDY

Traditionally in tribal society, land was not a commodity for sale. There was no concept of private property in land in most tribal communities. The process of separating the tribal people from their land originated in colonial times and their right to ownership of the forests they lived in was never recognized.

The introduction of the land revenue and money economy began this process. However, with capitalist development being stepped up, a continuous process of land alienation of the tribals has taken place. Their lands are taken away in the name of economic advancement of the country. But in return they receive landlessness, impoverishment and long-term degradation of the environment on which they totally depend. For almost two centuries now, tribal communities, like many other non-tribal peasants and forest-dwellers, have witnessed the collapse of their multiple relationships with the land, the forest and among themselves. Laws have been ineffective and inadequate. Through legal manipulation and other fraudulent means large-scale transfer of tribal land has taken place to non-tribal people. In such a crucial juncture, it is apt that adequate legislations be made for protecting and preventing the tribal land alienation/displacement/transfer for the improvement of their only source of sustainable livelihood. No doubt, a number of studies have been undertaken on the above aspect, but the present study would seek to find some of the gaps supposed to have been overlooked in other studies.

OBJECTIVES OF THE PRESENT STUDY

The present study aimed at the following objectives:

- to review and examine various land laws and regulations with special reference to alienation including acquisition, transactions and restoration of tribal lands.
- to assess the present status of immovable properties in the study area.
- to identify the reasons of land alienation and to study the methods of such alienation adopted in the scheduled areas.
- to study the role of Gram Sabha in preventing land alienation and restoration of land in scheduled areas.
- to assess the impact of land related laws especially alienation/transfer/displacement.
- to locate the gaps in implementation of various land related Acts.

METHODOLOGY AND SAMPLE DESIGN

The present study is an empirical study covering 2 districts, 6 blocks, 12 GPs and 24 villages and 3355 households. All households of sample villages were covered for the purpose. The methodology and sample design adopted in the present study is indicated in the Table 2.1.

SAMPLE

The study was carried out in two scheduled districts namely Koraput and Rayagada of Orissa. Three blocks from each district were selected based on (i) the incidence of land alienation and displacement in and around the areas, (ii) consultation of key stakeholders. Keeping this in view, Koraput, Semiliguda and Laxmipur blocks from Koraput district and Rayagada, Kolnara and Bisamcuttack blocks from Rayagada district were selected for the purpose.

The sampling was of a multi-stage purposive sampling keeping in view that the study areas have sufficiently large number of land alienation cases. From each block, two Panchayats were covered for the study. The selection of Gram Panchayats was done on the basis of suggestions given by the concerned officials, NGOs working on land issues and other key stakeholders. Baseline survey was conducted to identify the issue-based villages. Two villages from each selected Panchayats were selected keeping in mind that the villages may have had a high incidence of land alienation cases based on baseline survey report. All the physically existing households of the sample villages were covered in the study.

DETAILS OF SAMPLE: BLOCKS/GPs/VILLAGES

Methodology

Data was collected from both primary and secondary sources. Primary data includes (i) household survey, (ii) FGDs, (iii) semi-structured/informal interviews with the key stakeholders like elected Panchayat representatives, officials, activists, NGOs/CBOs, advocates and cross-section of the society, (iv) case studies, (v) field observations, and (vi) reality check/assessment

Table 2.1: Sample design

Description	**Select tribal districts of Orissa**					
Selected districts	**Koraput**			**Rayagada**		
Select blocks	**Koraput**	**Semiliguda**	**Laxmipur**	**Rayagada**	**Kolnara**	**Bisamcuttack**
Select GPs	Devighat Padampur	Rajput Pakjhola	Panchada Bhitaragada	Tadama Matikona	Therubali Dumuriguda	Bhatpur Chanchuraguda
Detail description of Sample Village	*Name of the villages surveyed and no. of HHs covered*	*Name of the villages surveyed and no. of HHs covered*	*Name of the villages surveyed and no. of HHs covered*	*Name of the villages surveyed and no. of HHs covered*	*Name of the villages surveyed and no. of HHs covered*	*Name of the villages surveyed and no. of HHs covered*
Sample Villages	Devighat (204)	Mukhibedei (232)	Panchada (569)	Tadama (143)	K.Dengasarigi (46)	Bhatpur (125)
	Dhaudapadar (132)	Sriramput (85)	Jhodipadar (42)	Nathama (91)	Gujulupadu (74)	Majhihaloma (36)
	Nighamaniguda (145)	Mania (331)	Bhitaragarh (143)	Matikona (109)	Bariguda (47)	Chanchuraguda (71)
	Machput (65)	Aligan (574)	Baragada (46)	Laximipuru (55)	Gunakhala (80)	Garadi (42)
Total HHs covered = 2439 in Koraput District + 916 in Rayagada district = 3355						

A. Primary source

Household Survey

All the existing households of the select villages were covered for the purpose. Pre-tested schedules/questionnaires were designed based on the information on family composition, income and expenditure, indebtedness, landholding particulars, details of land transfer, causes of transfer, methods of transfer, details of transferees, amount of transaction, problems in land regulations, role of Gram Sabha in preventing land alienation and restoration of lands to the tribal owners (*Refer Appendix-V*). A pilot survey was conducted prior to finalization of the schedules/questionnaire. Investigators were given an orientation on the concept of land related issues/ Acts/laws and techniques of data collection.

Focus Group Discussion (FGD)

FGDs in each selected village consisting of about 10 to 20 members representing the village were conducted. A semi-structured framework was designed based on the village information including land alienation and restoration issues in the local area. In each select village, 10 to 20 members comprising of both male and female were called upon to participate in Focus Group Discussion (FGD) representing the area. The team for conducting the FGDs comprised of three members. Out of them, one facilitator was meant to coordinate the discussions, another facilitator to record the discussions and the research officer was to interact with the participants according to predetermined framework.

Informal Interviews

Semi-structured informal interviews of the elected representatives of local bodies, officials, activists, NGOs/CBOs, and cross-sections of the society were conducted with regard to land alienation/displacement and restoration issues.

Case Studies

During the course of interaction with the villagers, key stakeholders, households and other target groups, some important case studies were recorded. The information was cross-checked and verified from local people and government records.

Field Observations

Land related issues were recorded during household survey and FGDs.

B. Secondary source

Secondary data were collected from revenue department, forest department records, NGOs associated in land issues/works, population and agriculture data, district statistical handbook, study reports, published and unpublished records/documents etc. Village level data was mainly based on information like village population, households, ethnic distribution, land holdings, land use pattern, land survey and settlement, village infrastructure, land alienation, migration due to land alienation, role of the Gram Sabha in preventing land alienation and effecting restoration.

Data were collected from RI office, district handbook and some from FGDs.

Apart from primary and secondary source, another tool was adopted i.e., reality check for the study purpose.

C. Reality check

Reality check is an act of verifying data as per government record in the field. Through this method, the gaps in and impact of the Acts can be identified. These can help government to undertake reforms at the policy level. This cross-check was conducted in both the select districts.

Limitations of the Study

- The present study suffers on account of the short duration of research period and therefore, was restricted to limited number of GPs and villages considering the number of issues which were required to be addressed.
- It was difficult to get the data on land alienation of the

past periods as most of the respondents of the households were minors or children at the time of alienation and therefore, were ignorant about the issues involved in alienation/restoration cases.

- Number of households did not match with the existing number as per the census data. In some cases, it was more and in some cases it was less. Hence, household data were collected from the existing households through household survey.
- In some cases, sons were found to be staying separately from their parents having their own income and expenditure. They earn their income by cultivating shared land, which have been transferred from their fathers/fore-fathers verbally. RoR (Patta) is recorded normally in the name of their father/eldest brother.
- Since they don't have *patta* in their own name as a result of verbal transfer through inheritance, it was considered that they had no land. This was necessary in order to avoid duplication in the research findings.
- Due to illiteracy, respondents could not give required information properly. Hence, the household data was likely to vary from actual. Of course, the researchers tried to collect authentic information as far as practicable.
- While calculating literacy rate in Rayagada district, children of 0-6 years were not included but in case of Koraput district, literacy rate was calculated including 0-6 years by mistake. So the comparison between the two districts cannot be reasonably and rationally valid.

3

Review of Literature

This chapter analyses the position of different studies undertaken in relation to various land related problems like survey, settlement and land reforms, land acquisition, land alienation, indebtedness, land rights of women, role of Gram Sabha and social audit ownership, control and management of MFP, common property resources, customary rights over forest land and impact of various laws in tribal areas.

Different institutions/organizations and social researchers have carried out a number of studies on different aspects of tribal land issues. A review of the following literature relating to different land issues was undertaken to find out the relevance of the present study on "Land Alienation and Entitlements in Select Tribal Districts of Orissa". The review of studies was done with a view to identify the gray areas, which could be taken up in the present study as far as practicable:

SURVEY, SETTLEMENT AND LAND REFORMS

The National Institute of Rural Development (NIRD) study team in their study on "Survey, Settlement and Land Reforms in Schedule-V Areas" in the year 2005 covered Andhra Pradesh, Madhya Pradesh, Orissa and tried to find out different forms of land ownership prior to survey operations, examined progress achieved in survey and settlement in recent years and reported the present status of survey and settlement of land records. The team briefly reviewed and examined various land laws and regulations as well as rights of the tribals over land including forest areas. They also tried to assess the impact of land reforms as well as survey and settlement operations. The team also tried to locate the legal and other constraints including gray areas in the existing laws and regulations constraining effective implementation of land reforms. Behuria in his book "Final Report on the Major Settlement Operation in Koraput District"(1965) also found in his study that no survey and settlement of pre-existing rights were taken up in the reserved and protected lands. In Kalahandi "dangara lands" during the pre-independence settlement, cultivation area of each Kondh *raiyat* (peasant), was noted in terms of seed capacity and accordingly rent was fixed in consultation or rather on persuasion with the *raiyat* (studied by *G.Das*, "Law Relating to Government Land and Lease Principles in Orissa" 1962). The period between 1960 and 1972 provided enough scope to the large land owners to transfer land in the name of their relatives while maintaining *de facto* control to escape land ceiling provisions (Mearns and Sinha, 1999). The study by Satapathy in his book "Land Reform Administration in Orissa" (1977) indicated that the progress of land reforms up to ceiling fixation and consolidation has been no doubt slow in some respects. But the significant achievement was that the tenants and the peasants are found to be gradually becoming more and more conscious of their rights. Hazzra in his "Land Reforms—Myths and Realities" (2006) focused actual scenario of West Bengal's much discussed land reform movement 'Operation Barga'. He presented a historical overview of land holding and cropping relations in both pre- and post-independent India tracing the roots of the land reform movement and related it to socio-historical and theoretical issues.

An attempt has been made by Mishra (1996) to study the question of land distribution among rural poor and its impact on their socio-economic conditions in the context of five districts (i.e Sitapur, Hardoi, Laxmipur Kheri, Baharich and Gonda) belonging to Uttar Pradesh. He has also made probe into the problems faced by allotees in the allotment and effective physical possession of land; the extent of land under legislation;

types of land allotted to them and their economic viability and the question of crop cultivation relating to output, income and employment generation. Whether or not the rural poor are aware of different poverty alleviation programmes is also taken into account because of their complementary relationship with the distribution of surplus land among the rural poor.

A study on"Land Rights and Ownership in Orissa-Land Study" taken up by UNDP (2008) discussed different land revenue assessment systems prevalent in India, particularly in Orissa both during the pre- and the post-colonial phases. It also discussed the provisions and impacts of existing land legislations of Orissa, besides the history and objectives of land reforms in the state.

LAND ACQUISITION

A study of NIRD in 2005 of Rajasthan, Chhatisgarh and Orissa on "Process of Acquisition of Land for Development of Mining Industry in Schedule-V Areas" examined the strength and weakness of PESA Act in empowering the Panchayati Raj Institutions with regard to the granting of land for mining and also awarding lease of major and minor minerals, the existence of any conflicts or contradictions between PESA and the earlier Acts on the subject as well as the procedures followed in the background of the PESA Act and the nature and extent of violations of PESA Act in these matters. ISED's study of 1998 on "Depriving the Underprivileged for Development" revealed (i) the socio-economic, cultural and environmental impact of such projects on the people - both those displaced and those likely to be displaced in future, (ii) reviewed the rehabilitation policies of different development projects of the state and examined the processes involved in their planning and the nature of their implementation. "Development, Displacement and Resettlement" of Verma, M.K. (2004) is a comprehensive sociological study of the processes and problems of development induced displacement and its ramifications on the socio-economic, cultural, psychological, political, ecological and environmental aspects of the project affected persons of two NTPC power projects.

Mr. Verma, besides identifying the impact of forced resettlement on the social structure values, beliefs, customs, and cultural essential conditions analyzed some specific problems like marginalisation, increased mobility, food insecurity and loss of access to common property resources. Most importantly, he ascertained the attitude of the land ousters towards the project and associated agencies. Last but not the least, it evaluated the adequacy and limitations of various rules and regulations governing the Land Acquisition Act and also resettlement and rehabilitation process and its understanding by the affected people.

The Status Report of UNDP (2008) critically examined development induced displacement till date and issues concerning compensation for such displacement. It also dealt with issues concerning fast growth of the mining and mineral based industrial sector and the associated displacements, establishment of 'Special Economic Zones' (SEZ), illegal land transfer, and encroachment of village commons. The study focused on various forms of land alienation caused due to forest conservation, commercial plantation and agriculture including contract farming.

LAND ALIENATION

There has been continual alienation of land of the marginalized, especially the tribals. No legislation has been able to put a halt to land alienation. Big landholders have been grabbing the land of tribals either by flexing muscles or by manipulation of land records joining hands with the land officials, or by lending money to tribals at exorbitant rates of interest. The study of NIRD (2005) on "Alienation of Tribal Lands in Schedule-V Areas" covered Andhra Pradesh, Madhya Pradesh and Orissa. The study focused on the incidence and magnitude of land alienation, the methods adopted for transfer of lands, its socio-economic and psychological impact on tribal people, particularly the women and the youth and an analysis on the legal strengths and weaknesses of the PESA Act and the role of Gram Sabha in preventing land alienation and restoring alienated tribal lands to the owners.

Dr. B.M. Bordoloi (1991) "Transfer and Alienation of Tribal Land in Assam with special reference to the Karbis of the Karbi, Anglong district made an attempt in his study on 'Transfer and Alienation of Tribal Land in Assam' to investigate as to how transfer and alienation of tribal lands are going on unabated in spite of the existence of legal provisions in the plane districts of Assam and also among the Karbis of the Karbi. The study reflects how adversely the transfer and alienation of lands has affected the various aspects of socio-economic life of the tribal people. Similarly alienation of tribal land due to establishment of industrial and irrigation projects and acquisition of land for public purposes are themselves problems of grave nature, which have resulted in causing sufferings of the affected families. The study tried to bring to light some hither to unknown facts in regard to the transfer and alienation of tribal lands in their proper perspectives. The extent of transfer and alienation, its 'modus operandi' and the consequences thereof are analyzed. After analyzing the causes, problems and consequences, suggestions are put forward for preventing transfer and alienation of tribal land. Behuria and Panigrahi (2001) in "A Report on functioning of the Fifth Schedule of the Constitution in the State of Orissa" tried to study the changes in the extent and nature of land alienation and strategies adopted by non-tribals and different agencies to alienate tribal people from their land resources and safeguards adopted to protect the interests of the tribal people on land. Approximately, 8550 acres of land has been sold through permission during the period 1957-1997 as per Regulation 2 of 1956 in Orissa (Panigrahi, 2001). Out of this, almost half was transferred in the four districts of Malkanigiri, Rayagada, Nowarangpur and Koraput (undivided Koraput district). The protective legislation regarding the problem of land alienation has been less effective. The reasons attributed by Yunus (1977) in his study "Orissa Chapter in Land Alienation and Restoration in Tribal Communities in India" are (i) lacuna in the laws, (ii) ignorance of the tribal people, (iii) complicated legal procedure involved in land restoration.

Patnaik, S. (2008) in UNDP Status Report, focused on specific provisions related to transfer and restoration of land, especially those that provide protection to tribal and scheduled caste owners of land.

INDEBTEDNESS

The study of NIRD on "Incidence of Indebtedness among the Tribals in Schedule-V Areas" covering Bihar, Madhya Pradesh and Orissa focused on very critical issues of incidence of indebtedness, the causative factors for indebtedness, examination of the relationship between money lending and land alienation and role of institutional agencies in extending credit to the scheduled tribes in the scheduled areas.

Viegas's study in four districts of Orissa found that scheduled tribes had lost almost 56 per cent of their private land, out of which 40 per cent was lost through debts and mortgages and rest 16 per cent through personal sales. (Viegas "Encroached and enslaved: alienation of tribal lands and its dynamics" 1991). A four-village study with a sample of 483 households found that 49 tribal households had mortgaged out a total of 44.5 acres of land. Of the 49 households 17 (35%) needed the money for marriage and funeral ceremonies and 23 per cent borrowed money for health related factors (Panigrahi, 2001).

A study on ' The problem of Land Alienation of the tribals in and around Ranchi' (Sinha in his book "Social Exclusion and Land Administration in Orissa"1968) stated that tribal indebtedness is one of the causes of land alienation. Some of the tribal landowners have been in heavy debt for one reason or the other and it was only to pay off their debt that they often sold away land for paltry sums of money. The administration also feels helpless in putting an end to alienation.

In a study on the tribals in Tamil Nadu, Raghava Rao in his study "Indebtedness in Tribal Society" (1991) has analyzed the extent of indebtedness and how it leads to alienation of land. The study reveals that out of 2631 tribal households, 54 per cent are under debt. He found that incidence of indebtedness had increased mainly because of heavy expenditure on domestic consumption, social and religious ceremonies, alcoholism, etc. and this had naturally led to misuse of loans advanced for productive purposes. This in turn, led to default in repayment due to lack of capacity which further accentuated leading to excess dependence on money-lenders, shop-keepers and non-tribal landowners.

WOMEN AND LAND RIGHTS

A study undertaken by ISED on "Women and Land Alienation in Jharkhand, Chattisgarh and Orissa" (2007) highlights the implications of the aggressive growth centric development on women's livelihood sources which are being taken away or expected to be taken in near future by the state to accommodate new ventures of the corporate sector. Besides, the study examined the increasing vulnerability of women in the wake of opening up of new opportunities as a result of aggressive investment. Patnaik, S. (2008) in his research study report on "Land Rights and Ownership in Orissa" specified about land rights of women across caste, class and religion.

LEGAL FRAMEWORKS

Government has been framing laws, Acts and policies to ensure rights and ownership of land to the poor and tribals. However, the reality is quite different. These Laws/ Acts remain as dead letters. Even though these are generally pro-people oriented, eventually, they turn out to be source of menace to these poor people. In actual practice, these facilitate the process for them to loose their livelihood and deprive them of their right from habitat.

The study on 'Impact of State Legislation' taken by the team on the 'Empowerment of Gram Sabha in Schedule-V areas' was to assess the impact of state legislation on the empowerment of Gram Sabha as well as to examine the levels of awareness of people and their consequent participation in the deliberations in Gram Sabha. In pursuance of this, the study made an analysis of relevant national and state legislations on the subject and tried to assess the degree of conformity of state legislation with the central PESA of 1996. The study findings reflected that various provisions of Central PESA empowering the Gram Sabha was limited. For example, the Gram Sabha is empowered by conferring the powers like (1) mandatory powers (a) power of prior approval of plans and projects for economic development by Gram Sabha, (b) power to issue certificate of utilization of funds by the Panchayat for the above programmes. (2) powers to be consulted in the acquisition of lands for public purpose and consequent displacement and rehabilitation of the affected people and (3) recommendatory powers for (a) grant of prospecting license or mining lease for minor minerals by auction (4) devolution of powers by the state government itself in various issues like ownership of MFP, control of money-lending and prevention of alienation of tribal lands. The study was undertaken in Madhya Pradesh, Gujarat, Rajasthan and Himachal Pradesh.

During the colonial period, the intermediary tenure holders were often at liberty to convert wastelands into cultivable holdings. This was based on explicit principle that wastelands belonged to the state though it allowed in principle that other persons who need to be compensated for state may claim rights of user can exercise complete control. Various tenancy and land revenue laws enacted after 1879, provided for leasing out of wastelands for agricultural and other purposes (Dash, 2004).

As per Orissa Prevention of Land Encroachment Act (1972), eviction or regularization of encroachments on government land depends on the basis of report of Revenue Inspector and proceedings initiated by the Tahsildar. Mearns and Sinha (1998) in their study "Social Exclusion and Land Administration in Orissa" found that there are few or no incentives for these officials to do so and frequently bribes are demanded or eviction threatened.

The Organization for Social Change and Rural Development's study used a sample of 477 ST beneficiaries/households drawn from 15 tribes spread over 3 scheduled districts of Orissa. The study analyzed the socio-economic profile of the sampled beneficiaries. Almost all the beneficiaries covered by the study were agriculturists. It would seem that the benefit of Regulation 2 of 1956 was going to tribal agriculturists and mainly the relatively well-to-do sections among the tribals.

Some of the major typologies affecting tribal access to land in Orissa have been focused in a study undertaken by Vasundhara team (2005) like loss of land settled with tribals through private transactions, lack of proper recording of occupation rights through survey and settlements, forests, forest land and tribal land, displacement through land acquisition for development projects, poor implementation of land regularization and distribution. The objectives of the study were (i) to identify typologies of factors constraining access of land to tribals and the consequences for tribal livelihoods and welfare, (ii) to explore causes and modalities of poor access of land to tribals, (iii) to explore constraints and opportunities to guarantee, restore and enforce tribal land rights, (iv) to explore positive and negative impacts of the existing legal and regulatory framework for tribal lands on the welfare and livelihoods of tribals, (v) to identify outcomes of laws and policies aimed at improving tenural security for tribal groups and possible policy changes required to improve access to land and tribal livelihoods.

Majhi, U.N. (2009), in his article "Tribal Land Alienation: Needs Radical Solution" highlights the issues of land alienation of tribal people in Orissa. He has given his views based on the studies taken from social scientists of 'Indian Social Institute', New Delhi and Institute for Social and Economic Change, Bangalore. He mentioned in his article that government is knowingly overlooking certain provisions of these Acts. In the 'Fifth Schedule Areas' where PESA is applicable, government is ignoring the existence of Gram Sabhas and signing MOU with the private companies. Government corporations should refrain from acting as brokers for granting acquired tribal land as lease to private companies. As per the recommendation of justice Vankatachaliah Committee, "Where interest of the tribal people are violated, application of sixth schedule is required for them for protection of their rights. In case all the committees, ombudsmen and commission fail to render justice, it is recommended for institution of tribal land tribunals at state and national level to adjudicate cases of land alienation, and rehabilitation and resettlement."

GRAM SABHA AND SOCIAL AUDIT

Another study taken up by the team of NIRD was on "Gram Sabha and Social Audit in Schedule-V Areas in Rajasthan, Maharasthtra and Madhya Pradesh". It focused on the analysis of social audit, role of the Gram Sabha on the framework of the provisions of PESA in the Schedule-V Areas. A modest attempt was made to examine the field scenario about the execution of PESA provision in order to know if empowerment of Gram Sabha has actually taken place.

OWNERSHIP CONTROL AND MANAGEMENT OF MFP

The study on ownership control and management of MFP in Schedule-V Area and the role of Gram Sabha/Village Panchayat was undertaken through NIRD in three states one each. One study was undertaken by RCDC, Bhubaneswar covering Andhra Pradesh, Maharastra and Rajasthan. All the three studies had the same objective i.e. to focus on the examination of nature and extent of empowerment conferred on Gram Sabha/Gram Panchayat by the PESA Act in matters of ownership, control and management of MFP, the underlying conflict/contradictions with the earlier Acts or regulations and the performance of Gram Sabha/GP in handling the new responsibility.

MANAGEMENT OF COMMON PROPERTY RESOURCES

Another study was undertaken by NIRD on 'Management of common property resources with special reference to minor water bodies in Schedule-V areas' covering Andhra Pradesh, Himachal Pradesh and Jharkhand. The study showed that uniformly across Schedule-V areas, the tribal people consider the village common government wasteland, water bodies and forest to be their common property resources. Access to these resources is ultimately linked to their survival and livelihood and forms the basis of their food security and health.

A recent empirical survey conducted in tribal dominated hill tracts of Orissa shows that on an average, a tribal household possesses 0.5 acres of land, which is less productive and susceptible to the damages caused by wild animals, pests and non-tribal exploitation. The tribal lands are normally located on the hill slope or on hilltop. In addition, whatever quantity of land tribals cultivate, a negligible percentage of that land is recorded against their names because lands located above (9) degree slope from the sea level is normally declared as government land. Tribal people use land patches for shifting cultivation and terrace cultivation, which are normally above (9) degree elevation. The proposal for recognizing the rights of the tribal people up to (30) degree slope of the 3 hills as proposed in case of Kashipur in Koraput district by Mohapatra, an anthropologist, was reportedly not accepted by the Government of Orissa.

TRIBAL CUSTOMS AND FOREST LAND

The study on "Customary mode of Dispute Resolution in ScheduleV areas" undertaken by the team of NIRD covered Madhaya Pradesh, Andhra Pradesh, and Orissa. The study was taken from selective Primitive Tribal Groups (PTGs) and major and minor tribes inhabitating the Schedule-V areas of Madhya Pradesh (now Chhatishgarh), Orissa and Andhra Pradesh focusing mainly on the structure and functioning of the traditional tribal and village committees and the mechanism of their customary mode of dispute resolution. The team observed that although the numerous tribal groups in the country displayed large diversity in their livelihood systems and cultural patterns, there is some fundamental commonality in the structure and functioning of the institutions in social control and dispute resolution. The study also showed that women were not normally involved in dispute resolution.

IMPACT OF FOREST LAWS

A study on "Impact of forest laws on livelihood systems with particular reference to tribal custom and natural resources in Schedule-V areas" undertaken by NIRD team which covered the Schedule-V areas of Bihar, Madhya Pradesh and Orissa, clearly explained about the existence of vital relationship between the forests and the traditional livelihood systems of tribal inhabitants of forests for the past several centuries. The study intended (i) to examine tribal dependence on forests and assessment of their current forest related needs like food (food gathering, hunting, fishing, farming etc.), shelter house building materials, firewood, fodder materials for household, cottage- industries tools and implements, cattle-grazing and of course, the cultural and religions practices of the tribals and

(ii) the dependence on MFP as a supplementary source of livelihood. Other aspects like impact of forest rules and attitude of forest officials, process of commercialization of NTFP, impact of forest policies on tribal women etc. was taken into consideration in the study. The study taken up by UNDP (2008) critically analysed the pros and cons of the recently passed Forest Rights Act, 2006.

Sahoo, A.C. (2009) in his article "The scheduled tribes and other traditional Forest Dwellers (Recognition of Forest Rights) Act, 2006: A critical view on forest policy, practice and operational issues" focused on the gaps in the Forest Rights Act and offered suggestions for effective implementation of the Act. Similarly, Patnaik, S. (2009) in his article "Forest and tribal livelihood" highlights the impact of the Act and offers valuable suggestions for better implementation of the Act.

Mohapatra, M.D. (2009), in her article, "The Forest Rights Act: An Instrument to ensure Forest Dwellers Rights" examined pros and cons of the Act. Sinha, S.P (2009) in his study on "The Problem of Land Alienation of the tribal in and around Ranchi" stated that tribal indebtedness is one of the causes of land alienation. Some of the tribal landowners have been in heavy debt for one reason or the other and it was only to pay off their debt that they often sold away land for paltry sums of money. The administrators also feel helpless in putting an end to land alienation. S.K. Chowdhury's study (2009) is based mainly on primary data in Garh-Tumilia village in Ramaghar block of Sundargarh district. Its aims were to study change in structure of landholdings and the extent to which *Zamindari* system was abolished in order to know persons affected by distribution of surplus ceiling lands and to assess efficiency in implementation by size and quality of land allotted and whether the land was effectively operated upon.

4

Profile of the Study Area

This chapter gives a total picture of the study area covering profile of Orissa with special reference to tribal situation in the state, profile of the study districts, study blocks and socio-economic profile of select villages.

PROFILE OF ORISSA

Orissa is located between 17 N and 22.3 latitudes and between 81.3 to 87.5 E longitudes and represents a State situated on the coast of Bay of Bengal, which is surrounded by Andhra Pradesh in the south, Bay of Bengal in the east, Chhatisgarh state in the west, West Bengal and Bihar states in the north. The geo-physical structure of the State is divided into four major zones depending upon the agro-climatic features. The state is divided into various productive zones by the department of agriculture, Government of Orissa. These are distributed in various geographical proportions. The state presently covers 1,55,707 sq. kms, which accounts for 4.74 per cent of the total land area of the country and geographically ranks in the 9th position among the Indian states (see Map-I). The administrative divisions of the state were formulated with 13 districts during 1948, again restructured into 30 districts in 1992, which are sub-divided into 59 sub-divisions, consisting of 314 community development Blocks, comprising of 51,124 number of villages. As far as the population strength of the state is concerned, it ranks 11th position and shares 3.74 per cent of India's population with a growth rate of 2.00 per cent per annum. The population of Orissa, which was 316.60 lakh in 1991, has increased to 368.05 lakh in 2001 exhibiting a decennial growth rate of 16.25 per cent as against 20.06 per cent in the previous decade and 23.86 per cent at the all-India level. The density of population increased from 203 per sq. km. in 1991 to 236 per sq. km. in 2001 which is lower than the all-India average of 313 per sq. km. The increase in the literacy rate from 49.10 per cent in 1991 to 63.08 per cent in 2001 was significant. The male and female literacy rates have gone up to 75.35 per cent and 50.51 per cent respectively in 2001. Since 1921, the state has experienced a progressive decline in the sex ratio of females. With regard to sex ratio while it was 1086 females per 1000 males in 1921, it has reduced to 927 females during 1991, but again increased to 972 as per 2001 census.

Orissa, situated in the heart of the eastern ghats, carries major rivers like the Mahanadi, Brahmani, Budhabalanaga, Subernarekha, Salandi, Kharswan, Rushikulya, Vansadhara, and Nagabali that drain into the Bay of Bengal. Rivers like Indravati, Kolab, Saberi, Machakunda and Sileru are the tributaries of river Godavari which are flowing through the undivided Koraput district, other rivers like Mahanadi, Brahamani and Subrnarekha have their origin outside the eastern ghats, but pass through the state of Orissa. A good number of multi-purpose dam projects on the river Mahanadi, Brahmani, Subrnarekha, Indravati, and Kolab have been constructed to control flood, provide irrigation and hydroelectric power.

According to the physical features and agro-climatic conditions, topographically, the state can be divided into four distinct zones. They are viz.

1. The Northern plateau, which accounts for 23 per cent of the total state and includes Mayurbhanj, Kenojhar, Sundargarh, and portions of Talcher of Anugul, Cuttack and Sambalpur districts. The hills ranges in these regions rise to an elevation of 600 to 900 m above the mean sea level.
2. The central table land consists of districts like Bolangir, southern part of Dhenkanal and part of Phulbani (now

Kandhamal) districts. The elevation spreads maximum to the point of 300 m to 800 m.

3. Eastern ghat region consists of hill ranges of main land of eastern ghats that spreads with extensive plateaus ranging from 300 to 450 m covering 36 per cent of the total land area of the state. It covers the districts like Koraput, Kalahandi (both undivided), Kandhmala and Gajapati.
4. The coastal plain covers 18 per cent of the total land area which runs from north to south and covers the districts of Balasore, Bhadrak, Cuttack, Kendrapada, Jagatsingh, part of Jajpur, Puri, part of Khurda and Ganjam.

The genesis of the soil in the state depends on the micro topographical situations and geomorphology of different types of land. The state has various types of soil like red-soil, laterite-soil, black-soil mixed-red and black type soil, brown soil, alluvial soil, coastal saline and sandy soil. The cultivable area according to the physiographic situation is generally divided into high lands which produces crops like maize, millets, upland paddy and *arhar*, the medium lands producing medium duration paddy; and the low lands generally produce late paddy of longer duration.

According to Agricultural Census, 2000-01 conducted by the Board of Revenue, there were 40.67 lakh operational holdings in Orissa with 50.81 lakh hectares of area. Small and marginal holdings accounted for 83.8 per cent with 53.12 per cent of total area respectively. Remaining 16.2 per cent of holdings belong to semi-medium, medium and large categories with 46.88 per cent of total area. Average size of operational holdings, which was 1.30 hectare in 1995-1996 census, declined to 1.25 hectare in 2000-01 census. Further, the agricultural census also reveals that out of the total of 40.67 lakh operational holdings, 17.99 lakh i.e. 44.2 per cent belong to scheduled castes and schedules tribes.

As regards the forest, there are two types of forest recognized by the Forest Act i.e. reserve forest and protected forest. The protected forest is divided as cultivable and non-cultivable. All the protected forest is covered under the control of revenue department. As a result, the state till date, has not made proper land use planning. The second dimension of the forest issue is the coverage of forests, which varies in total area and percentage to the land area. The difference still continues. The state has three types of major forests. They are : (i) the northern tropical semi-ever green forest, (ii) The northern tropical moist deciduous forests, and (iii) the northern tropical dry deciduous forests. The major forest produces in Orissa are timber, firewood, bamboo, minor forest produces, and kendu leaf. The rice potentiality and the availability of kendu leaf and timber in Orissa are reflected in the government revenue and forest records. The population growth in forest region, adoption of stringent forest policies by government and expansion of mega-development projects has considerably affected the forest resources of the State.

TRIBAL SITUATION IN ORISSA

Orissa has a landmass of 1,55,707 sq. km. with 3.68 crore population as per 2001 census. The scheduled tribes (ST) and scheduled castes (SC) population of the state are 81,45,081 and 60,82,063 respectively. The ST population constitutes 22.13 per cent of the total population of the state and 9.66 per cent of the total tribal population of the country. The SC population constitutes 16.53 per cent of the state and 3.65 per cent of the total SC population of the country. Both ST and SC together constitute 38.66 per cent of state's total population. Leaving apart the state of Nagaland, state of Orissa carries the highest percentage of tribal population in India. Considering the development index, the President of India, during 1956, declared 62 different tribal communities of Orissa as scheduled tribes out of which 13 are considered as Primitive Tribal Group (PTG) for special treatment. The scheduled tribes in Orissa speak as many as 74 dialects.

Although Mayurbhanj district is having the highest ST population (12.58 lakh) but the percentage of ST population to total population in Malkangiri district is the highest (57.43%). Similarly, Ganjam district is having the highest SC population (5.87 lakh) but the percentage of the SC population to total population in Sonepur district is the highest (23.62%). The sex ratio among ST and SC community is 1003 and 979 respectively.

It is observed that Koraput district has the highest proportion of families (83.61%) living below poverty line followed by Malkangiri with (81.88%), Boudh (80.2%), Deogarh (78.79%), Kandhamala (78.42%), Nuapada (78%), Mayurbhanja (77.7%) and Keonjhar (76.96%). All these districts are having predominance of tribal population. The districts having relatively higher proportion of families living below poverty line are always handicapped and more dependent on external financial resources.

Almost 44.21 per cent of the total land area in Orissa has been declared as scheduled area. Except the coastal belt, many of the districts of the state have been declared as either partially or fully scheduled area. The present scheduled area of the state covers six districts wholly and seven districts partially. The districts of Mayurbhanj, Sundargarh, Koraput, Nawarangpur, Malkangiri and Rayagada are wholly scheduled areas. Out of 314 Community Development Blocks in Orissa, 118 (37.3%) blocks are preponderantly occupied by the tribal communities, which are covered under Tribal Sub-Plan (TSP). The districts predominantly inhabited by tribal communities are Malkangiri (58.51%), Sundargarh (50.74%), Koraput (50.67%), whereas, the districts with a sizeable tribal population are Keonjhar (44.62%), Gajapati (47.88%), Jharsuguda (33.88%), Nuapada (35.95%), and Deogarh (33.31%). The major tribes of Orissa in terms of their numerical strength, to name a few, are the Kondhs, Gonds, Santals, Saoras, Bhuiyans, Parajas, Koyas, Oraons, Gadabas, Juangas and Mundas. Apart from these, there are also a good number of smaller tribal communities living in Orissa. Since the latest distribution of population tribe-wise was not available, an attempt has made to look into the census report of 1981, which reflects that the Chenchus, the smallest tribal community are representing only 39 persons. There are 15 tribal groups distributed in Orissa, each covering more than one lakh persons. When tribal communities like Santal, Gond, Munda, Ho, Birhor, Koya, Lodha, Kondha, Bhumija, Kharia and Oraons cut across the State boundaries and are found in the neighbouring states of Bihar, Chhatishgarh, Andhra Pradesh and West Bengal, the tribal communities like Juanga, Bonda and Didayee are found to be confined to the regions of their origin in the districts of Keonjhar and Malkangiri in Orissa.

Most of tribal communities of Orissa are of proto-Austroloid racial category. However, Mangoloi characteristics are also found among Gadabas and Saoras. The tribes of Orissa belong to three major linguistic divisions. They, by and large, maintain their linguistic identities and quite a sizeable of them belong to Mundari group of Austric language family and south Dravidian sub-group of Dravidian language family. In recent decades, few tribal groups like Santal, Ho, Saora, and Kui-Kandha have developed their own scripts. However, almost all these tribes have an un-documented script but well developed means of communication. A large number of tribals of Orissa speak Oriya and express various socio-cultural similarities with the neighbouring Oriyas in their life-way processes. Tribal societies due to certain common characteristics like animism, crude technology, concentrated habitation, use of a common dialect, subsistence economy etc. keep themselves distinguished from complex and advanced societies. They, by and large, manifest certain cultural features, which signify primitive levels in their socio-cultural milieu (Behura, 1988-1989: 1999).

PROFILE OF STUDY DISTRICTS

Koraput District

The district of Koraput derives its name from Koraput, the present headquarters of the district. In the medieval period, Nandapur was developed as the capital under the Silavamsi kings and sometimes under the kings of Solar dynasty. Vira Vikram Deo of Solar dynasty shifted his headquarters to Jeypore about the middle of the 17th century and this town was prospered as the capital. The British choose Koraput in 1870 for better health prospects. The origin of the name of

Koraput is obscure. There are several theories none of which is convincing. First of all, to Mr. R.C. Bell the name of the town is "Koraputte" or the hamlet of the nux vomica and it is derived presumably from a tree or trees that must, at one time, have been prominent near the site. But today not a single tree of nux vomica is to be found near about the town of Koraput and so the assumption of Mr. Bell is open to question. According to second theory, Koraput is corrupted form of 'Koraka Pentho'. Karaka literally means hailstone. It is also believed that one Khora Naiko laid the foundation of the village during the time of Nandapur king. He hailed probably from Ranapur and served under the Ranapur kings in the militia and for his faithful and meritorious services, he was permitted to establish this village which was named after him as Khoraput and later on the name has been abbreviated to Koraput. The district of Koraput was an unsurveyed tract and was carved out of the district of Vizagapatam in Madras with the creation of the province of Orissa on the 1st April, 1936. According to the government of India (Constitution of Orissa) Order, 1936, the Jeypore estates and so much of the Pottangi taluk as not included in that estate, was taken out from the Visakhapatnam district of Madras to be included in Orissa which formed the district of Koraput in Orissa. Thus, Koraput was first constituted as a district on the 1st April, 1936, when the province of Orissa came into existence. Before that it was a part of Madras presidency. It comprises of a greater part of the area formerly known as the agency tracts of Vizagapatam (now in Andhra Pradesh) which has an account of backwardness of the inhabitants administered by the collector with certain special powers in his capacity as agent to the Governor under the Ganjam and Vizagapatam Act of 1939.

Koraput is one of the southern-most districts of Orissa. Its boundaries extend in the north to Nawarangpur, Kalahandi and Rayagada districts, in the south to Andhra Pradesh, in the west to Malkangiri district of Orissa and Bastar district of Madhya Pradesh and in the east to Rayagada district and Andhra Pradesh. The geographical area of the district is 8.38 lakh ha which is 5.38 per cent of the state area. It is situated in between 18°-10° and 20° 10° north altitude, and 83°-20° east longitude. Physiographically the district is contiguous to the main land of eastern ghat. The district consists of two subdivisions, 14 C.D. blocks, 226 Gram Panchayats and 1997 revenue villages. (See Map-II) There is one Municipality (Jeypore) and three Notified Area Councils (Sunabeda, Koraput & Kotpad) in the district. The district is famous for Hindustan Aeronautics Limited (HAL), Sunabeda and National Aluminium Company (NALCO) at Damonjodi.

As per 2001 census, the total population of the district is 11,81,000 consisting of 9,82,000 rural population and 1,99,000 urban population. The female and male ratio is 999 females per 1000 males. The density of population is 134 per sq. km. Male literacy rate is 47.58 per cent and female literacy rate is 24.81 per cent in the district. Total SC population is 1,54,000 and ST populationis 5,86,000. Literacy rate in the district is 35.72 per cent. The literacy rate of STs is 18.68 per cent whereas SC literacy rate is 35.43 per cent. Forest area of this district is 3.23 per cent of the total forest area of the state. Out of total 5,70,000 workers in the district, 3,53,000 are main workers, 2,17,000 marginal workers, 1,85,000 cultivators and 2,29,000 are apicultural labourers.

RAYAGADA DISTRICT

The Rayagada district is known for its longest human history. The present district came functionally into existence on 2nd October in the year 1992. As the historical document goes, the region of Rayagada existed far back in the ancient times but except for a few copper plates, rock inscriptions and the coins there are no authentic written documents regarding the history of Rayagada district. However, it is evident from the coins and the inscriptions excavated that Rayagada was the center of attraction in all ages. In the 3rd century B.C. the, entire region of Rayagada was included in the Kalinga empire and King Ashoka was at the helm of administration. From the rock-edicts of Ashoka, it is seen that the history of Rayagada during the Ashokan regime is entirely printed. Bhojapur rock inscription in Brahmi of Rastrik king Mandhata adds some facts about the historical records of the Rayagada district. Kharavela, the only Aryan king ever ruled in Orissa defeated the Rastriks. Kharvela was ousted by the military campaign of Samudragupta. The Allahabad pillar inscription, framed by Harishena provides evidence to the socio-economic and cultural life of Rayagada during the reign of Samudragupta. During Gangavansi and Suryavansi kings, Rayagada was under Dadarnab Dev of Guderi Kataka. A rock inscription found at Pataleswar temple in Brahmi script depicts the era of the Ganga and the Surya kings in Rayagada. In the subsequent periods, many indigenous and foreign rulers colonized the region of Rayagada till significant part of Jeypore Zamindari, which later transformed into the princely state of the British till independence, survived as a sub-division of Jeypore Zamindary. From the perspective of the whole of Rayagada district it can be said that tribal communities are quite predominant here and comprise of 57.5 per cent of the total population. The typography of the district was such that the tribal people could easily retain their original tradition and ethnicity and live on their own whims and desires. Kondha tribe in the district is quite significant and stands apart among other tribal communities. In the city of Rayagada, the Kondha tribes are widely found. They are also concentrated in other parts of the district like Kashipur, Kalyansinghpur, Bisamcuttack and Muniguda as well. The Kondhas are a popular name amongst the tribes of Rayagada, with distinct features and characteristics. They belong to proto-Australoid racial group and are known for their practice of human sacrifice during British period. They follow a distinctive language namely "Kui". No written script was found. They are pure land-dwellers and have proved highly adapted to lush greenery of forest region. With passage of time, these tribes of Rayagada seem to have developed and progressed in various fields like education, irrigational opportunities, plantation facilities, advancement in medicinal plants etc.

Rayagada district comprises of two subdivisions i.e. Rayagada and Gunupur. The district occupies the southern part of the state bordering the state of Andhra Pradesh. The district is located between *82° 54′ to 84° 02′* East longitude and *19° 00′ to 19° 58′* North latitude. The district has an area of 7073 sq. km. It is surrounded by Gajpati district on the east, Andhra Pradesh on the south, Koraput district to the South-West, Kalahandi district to the West and North-West and Kandhamal district to the North-East. The district has four Tahasils, eleven blocks, one municipality and two N.A.Cs and two census towns (See Map-III). The district comprises of 171 Gram Panchayats, 2,667 villages of which 200 are uninhabited and 2467 are inhabited. The district has four Assembly constituencies. The highest elevations of 1100 to 1500 meter is found in the block of Kashipur which is contiguous in nature from west to east up to Liliguma R.F. Rivers like Naghulu, Golagarhi, Barha have their origin from these mountain ranges and flow towards Kalahandi district. River Patagarha, the main tributary of the Nagavali river has its origin from this region too. Kalyan-Singhpur, Muniguda and Ambadola villages are located over the hills whose heights vary from 1,100 to 1,500 meter above mean sea level. Most of the hill-tops are covered with deciduous forest of reserve category. According to 2001 census, Rayagada has shown a population count of 8,32,019 out of which male population constitutes 51 per cent. In comparison, female populace is less and thus constitutes 49 per cent. Children who are below 6 years of age constitute 12 per cent of the total population. The literacy rate of Rayagada is quite low as compared to the national average. The average literacy rates of Rayagada and India compare at 63.08 per cent and 36.15 per cent respectively.

PROFILE OF STUDY BLOCKS OF KORAPUT AND RAYAGADA DISTRICT

As mentioned earlier, the Blocks and Gram Panchayats from each district were selected for study based on the key stakeholders consultation and field visits. Six blocks namely Koraput, Semiliguda and Laxmipur from Koraput district and Rayagada, Kolnara and Bisamcuttack from Rayagada district were selected for the purpose. (See Map-IV, V, VI, VII, VIII, IX) Demographic profile of select blocks in two districts is shown in Table 4.1.

Table 4.1: Demographic profile of select blocks of Koraput and Rayagada district

Name of the District	Name of the Block	Area in Sq. Km.	No of Villages	No. of HHs	Total Population	Density	Male Population	Female Population	Sex-Ratio
Koraput	Laxmipur	310.33	104	13600	55268	178.09	27452 (49.67)	27816 (50.33)	1013
	Semiliguda	314.65	85	14026	55537	176.5	27373 (49.29)	28164 (50.71)	1029
	Koraput	420.89	92	17083	57572	136.79	28931 (50.25)	28641 (49.75)	990
Rayagada	Rayagada	530.70	293	24913	112330	176.27	54776 (48.76)	57554 (51.24)	1051
	Bissam-cuttack	620.38	307	19039	82774	121.16	40155 (48.51)	42619 (51.49)	1061
	Kolnara	366.56	199	15148	66670	186.94	32483 (48.72)	34187 (51.28)	1052
	Total	2563.51	1080	103809	430151	167.80	211170 (49.09)	218981 (51.91)	1037

Note : Figures in the bracket indicate percentage to the total

Source : Census of India 2001

According to the census report of 2001, the total population of the study blocks is 4, 30,151 out of which 2,11,170 (49.09%) are males and 2,18,981 (51.91%) females. Thus the female population is more than the male population in the study blocks.

Table 4.2: Ethnic distribution of population of select blocks of Koraput and Rayagada district

Name of the District	Name of the Block	Total Population	Total SC Population	Total ST Population	Others
Koraput	Laxmipur	55268	7383 (13.36)	37951 (68.67)	9934 (17.97)
	Semiliguda	55537	4996 (9.0)	30364 (54.67)	20177 (36.33)
	Koraput	57572	8496 (14.76)	24460 (42.49)	24616 (42.75)
Rayagada	Rayagada	112330	11483 (10.22)	71417 (63.58)	29430 (26.2)
	Bissamcuttack	82774	14083 (17.01)	50360 (60.84)	18331 (22.15)
	Kolnara	66670	5901 (8.85)	46612 (69.92)	14157 (21.23)
	Total	430151	52342 (12.17)	261164 (60.72)	116645 (27.11)

Note : Figures in the bracket indicate percentage to the total
Source : Census of India 2001

While examining the position of ethnic distribution of population in select blocks as per 2001 census (shown in Table 4.2) it was found that out of 4,30,151 total population, more than 60% of population belonged to scheduled tribes, 12.17% to scheduled castes and 27.11% to other castes. Block-wise data reveals that except Koraput block, all select blocks almost have the same population trend.

Table 4.3: Sex-wise distribution of population of select blocks of Koraput and Rayagada district

Name of the District	Name of the Block	Total SC Population	Male SC Population	Female SC Population	Total ST Population	Male ST Population	Female ST Population
Koraput	Laxmipur	7383	3679 (49.83)	3704 (50.17)	37951	18745 (49.39)	19206 (50.61)
	Semiliguda	4996	2480 (49.64)	2516 (50.36)	30364	14980 (49.33)	15384 (50.67)
	Koraput	8496	4182 (49.22)	4314 (50.78)	24460	11947 (48.84)	12513 (51.16)
Rayagada	Rayagada	11483	5673 (49.4)	5810 (50.6)	71417	34225 (47.92)	37192 (52.08)
	Bissamcuttack	14083	6945 (49.31)	7138 (50.69)	50360	24046 (47.75)	26314 (52.25)
	Kolnara	5901	2947 (49.94)	2954 (50.06)	46612	22392 (48.04)	24220 (51.96)
	Total	52342	25906 (49.49)	26436 (50.51)	261164	126335 (48.37)	134829 (51.63)

Note : *Figures in the bracket indicate percentage to the total*
Source : *Census of India 2001*

Sex-wise and caste-wise distribution of population data is revealed in Table 4.3 above. In both the castes, female population is more than male population. Out of total 52342 SC population, 50.51 per cent were female and out of 26436 ST population, 51.63 per cent respectively belong to the female population whereas the male population of both the SC and ST constitute 49.49 and 48.37 per cent respectively.

Table 4.4: Literacy rate of select blocks of Koraput and Rayagada district

Name of the District	Name of the Block	Total Population	Literates			Total literacy rate (per centage)
			Male	Female	Total	
Koraput	Laxmipur	55268	8886 (70.41)	3735 (29.59)	12621	22.84
	Semiliguda	55537	8545 (77.11)	2536 (22.89)	11081	19.95
	Koraput	57572	12407 (71.42)	4966 (28.58)	17373	30.18
Rayagada	Rayagada	112330	13569 (67.05)	6669 (32.95)	20238	18.02
	Bissamcuttack	82774	11204 (69.82)	4843 (30.18)	16047	19.39
	Kolnara	66670	12535 (66.12)	6422 (33.88)	18957	28.43
	Total	430151	67146	29171	96317	22.39

Note : Figures in the bracket indicate percentage to the total
Source : Census of India 2001

One of the reasons for the backwardness and poverty in tribal areas is illiteracy. The present study made an attempt to look into the literacy status as per 2001 census report. The block-wise level of literacy varies from a minimum of 18.02 per cent in Rayagada block to a maximum of 30.18 per cent in Koraput block (indicated in Table 4.4). However, the female literacy rate is very low in all the studied blocks in comparison to the male literacy rate. On the whole, out of 96317 total literates, male literacy constitutes 69.71 and female literacy is 30.29. In Koraput district, highest male literacy was noticed in Semiliguda block and in Rayagada district highest rate was found in Bissamcuttack block.

Occupational structure is a prime indicator of development of an economy. It determines the employment status, nature of employment as well as economic position and standard of living of persons. It will also be helpful for the planners to develop strategies for area-wise development.

Table 4.5 depicts the data on occupational pattern of study districts as per 2001 census. It shows that proportion of total workers to total population is 48.31 and 48.03 in Koraput and Rayagada districts respectively.

Table 4.5 reveals that the percentage of (62.60) main workers in Rayagada district is more than Koraput district (61.95). Out of the total main workers, 42.04 per cent are cultivators and 22.63 per cent are agricultural labourers in Koraput district while in Rayagada district 38.7 per cent are cultivators and 30.58 per cent are agricultural workers. Thus, the percentage of cultivators was found to be more in Koraput district while the percentage of agricultural labourers was more in Rayagada district. The marginal workers constitute 38.05 per cent to total workers in Koraput district while there were 37.40 per cent of marginal workers in Rayagada District (Shown in Fig. 4.1 and 4.2). This scenario would draw the attention of the planners to think how to face the challenges of employment in providing timely support to the people in this regard.

Table 4.5: Occupational pattern of Koraput and Rayagada district

Occupational Pattern	Total Persons		Description	Percentage	
	Koraput District	Rayagada District		Koraput District	Rayagada District
Main Workers*	353367	249909	Proportion of main workers to total workers	61.95	62.60
Cultivators	148578	95383	Proportion of cultivators to main workers	42.04	38.17
Agricultural Labourers	79975	76428	Proportion of agricultural labourers to main workers	22.63	30.58
Household Industry	7464	4666	Proportion of household industry to main workers	2.11	1.87
Other Workers	117350	73432	Proportion of other workers to main workers	33.21	29.38
Marginal Workers	217068	149275	Proportion of marginal workers to total workers	38.05	37.40
Total Workers	570435	399184	Proportion of total workers to total population	48.31	48.03
Total Population	1180637	831109	–	–	–

**Source* : Economic Survey 2007-08

*Main workers include cultivators, Agricultural Labourers, Household Industry and Other Workers.

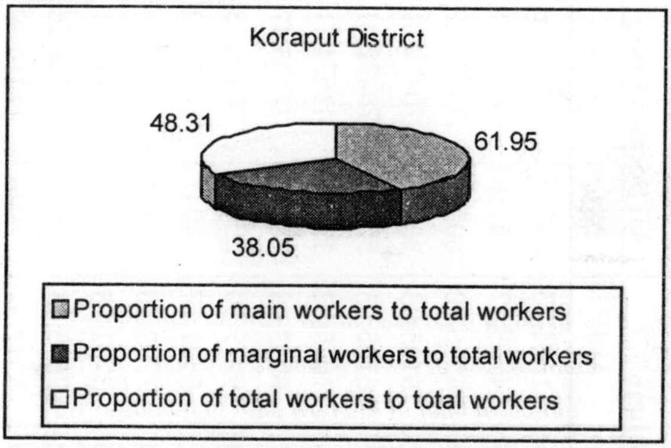

Fig. 4.1: Occupational pattern in Koraput district

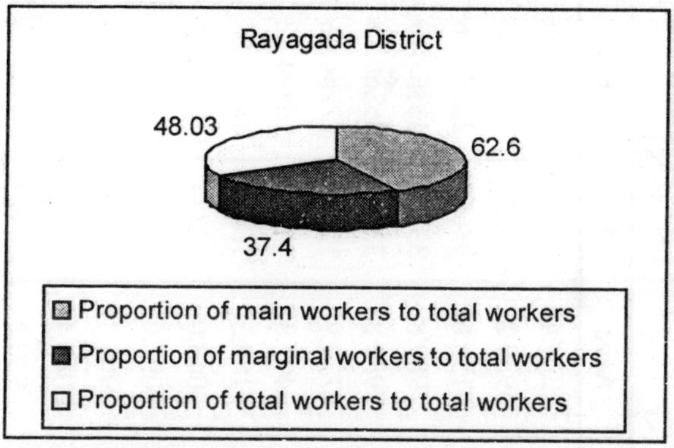

Fig. 4.2: Occupational pattern in Rayagada district

The land holding pattern symbolizes the economic position of an area. By this, area planning will be successfully drawn by putting emphasis on different types and categories of people. This is reflected in the Table 4.6:

As per Agricultural Census 2000-01 (revealed in Table 4.6 above for Koraput district and Table 4.7 for Rayagada district) the size of land holding is categorized into 5 groups viz. (i) marginal up to 1 hectare, (ii) Small up to 2 hectare, (iii) semi-medium 2-4 hectare, (iv) medium 4-10 hectare and (v) large above 10 hectare whereas the land holders are classified on the basis of different social groups viz. institutional, scheduled caste, scheduled tribes and others.

Table 4.6: Estimated number and area of operational holdings in Koraput district

Size Class	Institutional		SC		ST		Others		All Social Group	
(In hac)	No.	Area	No.	Area	No.	Area	No.	Area	No.	Area
Below 0.5	22	4.40	4277	1219.40	14323	4586.33	5515	1681.33	24137	7491.46
(0.5-1.0)	5	3.89	5433	3939.22	24335	17767.23	6385	4555.47	36158	26265.81
Marginal	27	8.29	9710	5158.62	38658	22353.56	11900	6236.80	60295	33757.27
(1.0-2.0)	6	7.86	5715	7989.83	33018	46369.19	11442	16105.93	50181	70472.81
Small	6	7.86	5715	7989.83	33018	46369.19	11442	16105.93	50181	70472.81
(2.0-3.0)	0	0.00	1950	4774.93	11502	27754.46	4526	11045.32	17978	43574.71
(3.0-4.0)	0	0.00	714	2407.37	4918	16694.12	2387	8233.98	8019	27335.47
Semi-Medium	0	0.00	2664	7182.30	16420	44448.58	6913	19279.30	25997	70910.18
(4.0-5.0)	2	8.72	448	2043.09	2344	10404.94	1247	5521.17	4041	17977.92
(5.0-7.5)	2	12.39	519	3097.08	2304	14035.68	1289	7792.48	4114	24937.63
(7.5-10.0)	5	41.73	100	903.27	1002	8715.83	436	3781.78	1543	13442.61
Medium	9	62.84	1067	6043.44	5650	33156.45	2972	17095.43	9698	56358.16
(10.0-20.0)	15	207.98	80	930.11	419	4959.99	432	5312.31	946	11410.39
20 & Above	46	5650.11	5	125.46	45	2197.03	49	1261.74	145	9234.34
Large	61	5858.09	85	1055.57	464	7157.02	481	6574.05	1091	20644.73
All Classes	103	5937.08	19241	27429.76	94210	153484.80	33708	65291.51	147262	252143.15

Source: Agricultural Census 2000-01

Table 4.7: Estimated number and area of operational holdings in Rayagada district

Size Class	Institutional		SC		ST		Others		All Social Group	
(In hac)	No.	Area	No.	Area	No.	Area	No.	Area	No.	Area
Below 0.5	2	0.80	1735	537.11	13284	4339.62	2308	725.31	17329	5602.84
(0.5-1.0)	0	0.00	2780	1957.37	22054	16521.43	4102	2881.93	28936	21360.73
Marginal	2	0.80	4515	2494.48	35338	20861.05	6410	3607.24	46265	26963.57
(1.0-2.0)	7	9.86	2590	3385.39	23236	31789.60	5430	7257.50	31263	42442.35
Small	7	9.86	2590	3385.39	23236	31789.60	5430	7257.50	31263	42442.35
(2.0-3.0)	7	14.52	885	2059.78	8214	19648.71	2380	5791.34	11486	27514.35
(3.0-4.0)	2	7.30	330	1120.19	3418	11658.16	1050	3584.31	4800	16369.96
Semi-Medium	9	21.82	1215	3179.97	11632	31306.87	3430	9375.65	16286	43884.31
(4.0-5.0)	0	0.00	120	528.69	1729	7602.40	774	3412.10	2623	11543.19
(5.0-7.5)	6	35.55	100	574.67	1540	9105.62	770	4550.40	2416	14266.24
(7.5-10.0)	0	0.00	50	410.28	330	2779.96	295	2506.34	675	5696.58
Medium	6	35.55	270	1513.64	3599	19487.98	1839	10468.84	5714	31506.01
(10.0-20.0)	1	12.00	20	246.71	239	2986.67	245	3207.24	505	6452.62
20 & Above	8	358.66	5	100.36	10	243.68	40	1036.13	63	1738.83
Large	9	370.66	25	347.07	249	3230.35	285	4243.37	568	8191.45
All Classes	33	438.69	8615	10920.55	74054	106675.85	17394	34952.60	100096	152987.69

Source: Agricultural Census 2000-01

In Koraput district, out of 147262 total landholders, 103 holdings belonged to Institutions, 19241 number of holdings belonged to SCs, 94210 to STs and 33708 holdings were possessed by OBCs. Similarly, in Rayagada district, out of 100096 total holdings, different institutions possessed 33 land-holdings, 8615 belonged to SCs, 74054 landholdings was possessed by STs and 17394 were OBCs. As regards the class of operational holding of HHs, the highest number of landholders belongs to marginal class in both the districts.

To assess the economic condition of an area it is important to know how best the land of an area is utilized. As per 2001 census, broadly the land use pattern is classified into forest area, misc. tree crops and groves not included in net area sown, permanent pasture and other grazing lands, cultivable waste, land put to non-agricultural uses, barren and uncultivable lands, current fallows, other fallow lands and net area. Table 4.8 depicts the data on land utilization pattern in the select blocks of two districts.

SOCIO-ECONOMIC AND DEMOGRAPHIC PROFILE OF STUDY VILLAGES

The data were collected both from the primary and secondary sources. The study was conducted in the districts of Koraput and Rayagada during 2008-2010. These districts have the largest concentration of tribal population of the state. Altogether 3355 households covering 24 villages spread over 6 blocks were interviewed by selecting 2 villages from each block and 3 Blocks from each district. Explanatory household and village schedules were used to collect relevant data. Household data pertaining to land related information and socio-economic conditions were administered in the households of select villages. Review of various research studies shows that land alienation is found to be more in tribal areas because of their poor socio-economic conditions and illiteracy. Socio-economic profile can indicate the standard of living and social status. Keeping this in view, apart from collecting land related information from the HHs, the present study tried to assess the socio-economic conditions of HHs and infrastructure available in the study villages of both the districts.

Table: 4.8: Land utilization pattern in select blocks of Koraput and Rayagada district (area in hac)

Block	Forest area	Misc. tree crops and groves not included in net area sown	Permanent pasture and other grazing lands	Cultivable waste	Land put to non-agricultural uses	Barren and un-cultivable lands	Current fallows	Other fallows	Net area sown
Laxmipur	1014	1369	758	509	1583	11762	6630	297	7049
Semiliguda	1244	718	536	1270	490	12783	2881	1782	10694
Koraput	3017	3445	1058	121	1906	5150	4493	1655	10669
Bisamcuttack	9739	695	1013	562	2240	23641	6715	2319	16277
Kolnara	4449	573	540	982	2442	16253	2643	1769	12115
Rayagada	8315	345	692	620	3216	24682	3511	2663	12298

Source: District Statistical Handbook, Orissa 2001.

Table 4.9: Demographic profile of select villages of Koraput district as per 2001 census

Name of the Block	Name of the village	Area (in Sq. Km)	Total HHs	Total Population	Male	Female
Koraput	Devighat	331	256	1061	571	490
	Dhaudapadar	273	176	652	316	336
	Nighamaniguda	107	155	729	360	369
	Machput	189	65	220	99	121
Semiliguda	Mukhibedei	669	289	1111	559	552
	Sriramput	172	51	200	98	102
	Mania	890	331	1318	684	634
	Aligan	1914	574	2073	1054	1019
Laxmipur	Panchada	1895	675	2727	1414	1313
	Jhodipadar	101	36	137	69	68
	Bhitaragada	295	134	509	244	265
	Baraguda	135	51	176	81	95

Source: Census of India 2001

Table 4.9 describes the demographic profile of the studied villages of Koraput district (as per 2001 census). Among the twelve studied villages of Koraput district, seven villages namely Dhaudapadar, Nighamaniguda, Machput, Sriramput, Panchada, Jhodipadar, Bhitaragada and Baraguda are female dominated and the other five villages Devighat, Mukhibedei, Mania, Aligan and Jhodipadar had male domination. As a whole, the studied villages of Koraput district were found to be female dominated. Aligaon village has the highest area and Jhodipadar village the lowest area in the study district of Koraput.

It is revealed from the demographic profile of Rayagada district (Table 4.10) that out of twelve studied villages, three villages viz. Matikona, Laxmipur of Rayagada block and Garadi village of Bissamcuttack block are male dominated villages. But in Garadi village the difference between male and female number was only one. It was also found that in all the studied villages of Kolanara block, female population is predominantly more.

The data on infrastructure availability were assessed by conducting FGD in each select village and information was collected accordingly.

Table.4.10: Demographic profile of select villages of Rayagada district as per 2001 census

Name of the Block	Name of the village	Area (in Sq. Km)	Total HHs	Total Population	Male	Female
Rayagada	Tadama	520	214	913	441	472
	Nathama	268	113	496	244	252
	Matikona	300	103	477	251	226
	Laxmipuru	NA	NA	234	121	113
Bissam-	Bhatpur	802	147	625	288	337
cuttack	Majhihaloma	826	35	168	76	92
	Chanchuraguda	64	73	315	152	163
	Garadi	74	33	143	72	71
Kolnara	Dengasingi	418	105	456	219	237
	Gujalpadu	92	71	295	140	155
	Bariguda	NA	NA	NA	NA	NA
	Gunakhal	624	135	650	313	337

Source: Census of India 2001

INFRASTRUCTURE AVAILABILITY IN STUDY AREA

The living condition of villagers depends on the degree of access to infrastructure facilities. Infrastructure refers mainly to infrastructure and institutional facilities like schools, post offices and telegraphs, telephones, concrete roads, electricity, police stations, fair-price shops, financial institutions, health-care services, drinking water etc. While enquiring about availability of infrastructure facilities in the study villages of Koraput district, it was found that in every village there is at least one primary school. All the villages have post offices within a radius of 5 kms. Of course, Devighat village itself has a post office. Telephone facility is available only in some villages. In some villages people have access to mobile phones. Except some locations of Panchada, all the villages are having concrete roads. This has been possible due to implementation of different schemes of Ministry of Rural Development, Government of India and State government. Electricity facility is not available in Dhaudapadar, Srirampur, Panchada, Jhodipadar and Baraguda villages. In rest of the villages, about 10-30 HHs have been electrified. Police Station does not exist in any of the villages. It is located at places far off from the villages say, at a distance of five to fifteen kms. Fair-price shops (Ration Shop) are available in every GP, which is located almost 2-5 kms away from each village except the GP headquartered villages. Financial Institutions like banks are situated at a distance of almost 7-15 kms from the villages. People have not easy access to health-care facilities. They have to cover about 15 kms to get health services in some cases. Of course, mobile health services are being provided to these villages. In every village, tube-wells have been provided. But maximum tube-wells are reportedly not functioning. NGOs have some sort of intervention in every study village except village Baraguda. At least 5 SHGs are found in every study village. RI offices are functioning in the respective circles. These are normally located in a radius of 5-15 kms. Forest offices are located mostly at places where blocks are situated but not nearer to the villages. Veterinary offices are mostly located at a distance of maximum 15 kms away from the villages. Weekly markets (Haats) are also not nearer to these villages. Hardly any village has clubs/ community centers of its own. On the whole, it was found that there is severe problem of drinking water in some of the villages. For instance, during the course of FGD in Machhput village, some of the villagers became irritated and aggressive while answering questions about land related issues indicating that these things are not of any use to them as their priority problem is drinking water. In fact, this was their first priority.

Like-wise, while enquiring about availability of infrastructure facilities in the study villages of Rayagada district, it was found that every village is accessible and well-connected with the block headquarters except for the village Majhihalma. This village is isolated and having no connecting road to it. People are using both sides of the railway track as their regular path to enter into the village. Except Majhihalma, all other villages are having concrete roads. This has been possible due to implementation of different schemes of Ministry of Rural Development,

Government of India and the state government. There is at least one primary school in each village except village Bariguda of Therubali GP. All the villages have post offices located within a radius of 5 to 9 kms. No public telephone facility is available in the study villages. But in every village 10 to 15 mobile phones are in use. Electricity facility is not available in Majhihalma. In the rest of the villages, about 10-30 HHs have electrification in their respective houses. Police station does not exist in any of the villages. It is far off from the villages say at a distance of five to fifteen kms. Fair-price shops (Ration Shop) are available in every GP, which is located almost 2-5 kms away from each village expect the GP headquartered villages. Financial institutions like banks are situated almost 7-15 kms from the villages. People have no easy access to health-care facilities. They have to go about 10 to 15 kms to avail medical facilities. In every village, tube-wells have been provided but most of the tube-wells are not functioning. In every village, SHGs are functioning actively as the catalyst for socio-economic development of the village. RI offices are found in the respective circles. These are normally located in a radius of 5-15 kms. Forest offices are located mostly at places, where blocks are situated not nearer to the villages. Veterinary offices are mostly located at a distance of maximum 15 kms away from the villages. Weekly markets (Haats) are also not located nearer to these villages. Hardly any village has clubs/community centers of its own.

DEMOGRAPHIC STRUCTURE

The composition of population reflects the pattern of social life and overall social relationship. Basically, the inputs and outputs of the society towards the day-to-day life are determined by the size of population. Besides size of the population, the differences in attitudinal and cultural life along with social life influence the developmental process. An attempt was made through this study to know the population of the studied villages of Koraput and Rayagada districts in detail.

Table 4.11(a): Population of study villages of Koraput district

Name of the village	No. of existing household surveyed	Number of persons	No. of male	No. of female	Average size of family
Devighat	204	844	404 (47.87)	440 (52.13)	4.1
Dhaudapadar	132	490	236 (48.16)	254 (51.84)	3.7
Machput	66	264	116 (43.94)	148 (56.06)	4.0
Nighamaniguda	145	662	301 (45.47)	361 (54.53)	4.6
Bhitaragada	143	585	294 (50.26)	291 (49.74)	4.1
Baraguda	46	196	102 (52.04)	94 (47.96)	4.3
Panchada	569	2366	1186 (50.13)	1180 (49.87)	4.1
Jhodipadar	42	163	85 (52.15)	78 (47.85)	3.9
Mukhibedei	232	1029	501 (48.69)	528 (51.31)	4.4
Sriramput	85	316	153 (48.42)	163 (51.58)	3.7
Mania	331	1425	695 (48.77)	730 (51.23)	4.3
Aligan	444	1793	882 (49.19)	911 (50.81)	4.0
Total	2439	10133	4955 (48.9)	5178 (51.1)	4.1

Note: Figures in the bracket indicate percentage to the total
Source: Field Survey 2008-2009

Table 4.11(a) presents the demographic structure of the study villages of Koraput district. It is revealed from the table that proportion of female members is more than the proportion of male members in the study villages As found in the study, total number of family members is 10133 out of which total male population is 4955 (48.9 %) and female population is 5178 (51.1%). Among the 12 villages surveyed, Jhodipadar has the highest percentage of male population i.e. 52.15 whereas female population is maximum in Machput village i.e. 43.94 per cent. Among the study villages, the proportion of women members is more than the male members except in 4 villages i.e. Bhitaragada, Baraguda, Panchada, Jhodipadar. Average size of the family in the study villages is 4.1, which implies that the size of the tribal family is not that large. This may be due to high mortality rate or low fertility rate among the tribes.

Table 4.11(b): Population of study villages of Rayagada district

Name of the village	No. of existing household surveyed	Number of persons	No. of male	No. of female	Average size of family
Garadi	42	187	84 (44.92)	103 (55.08)	4.4
Chanchadaguda	71	323	146 (45.20)	177 (54.80)	4.5
Bhatapur	125	571	263 (46.06)	308 (53.94)	4.6
Majhihalma	36	160	69 (43.13)	91 (56.87)	4.4
Gujalpadu	74	376	172 (45.74)	204 (54.26)	5.1
K. Dengasargi	43	210	101 (48.10)	109 (51.90)	4.9
Gunakhal	80	391	187 (47.83)	204 (52.17)	4.9
Bariguda	47	213	110 (51.64)	103 (48.36)	4.5
Matikona	109	434	195 (44.93)	239 (55.07)	4.0
Laxmipur	55	162	88 (54.32)	74 (45.68)	2.9
Nathama	91	449	199 (44.32)	250 (55.68)	4.9
Tadama	143	644	309 (47.99)	335 (55.01)	4.5
Total	916	4120	1923 (46.67)	2197 (53.33)	4.5

Note: Figures in the bracket indicate percentage to the total
Source: Field Survey 2009-2010

The figures in the Table 4.11 (b) reflect the population of the sampled villages. The total population of the twelve villages is 4120 out of which 1923 (46.67%) are male and 2197 (53.33%) females. It also reveals that, except two villages viz. Bariguda and Laxmipuru, all other studied villages are female dominated. The average family size of the selected population is 4.5 for the sample. There is a slight variation of family size and this is also inevitable. However, the average size of the family is lowest in Laxmipur (2.9) and highest in Gujalpadu (5.1). This confirms the trend which is shifting towards nuclear family system due to various socio-economic factors.

AGE OF THE HOUSEHOLD

Decision-making is a crucial factor in each and every sector of social life. Likewise, in household affairs, generally important decisions are being taken by the head of the households. The present study tried to find the average age of the heads of households.

Table 4.12: Average age of head of household of study districts

Koraput	**District**	**Rayagada District**	
Name of the Village	**Average age of Head of Household**	**Name of the Village**	**Average age of Head of Household**
Devighat	39	Garadi	36
Dhaudapadar	41	Chanchadaguda	35
Machhput	43	Bhatapur	35
Nighamaniguda	44	Majhihalma	36
Bhitaragada	38	Gujalpadu	36
Baraguda	48	K. Dengasargi	35
Panchada	41	Gunakhal	35
Jhodipadar	41	Bariguda	35
Mukhibedei	40	Matikona	36
Sriramput	40	Laxmipur	36
Mania	40	Nathama	35
Aligan	40	Tadama	35
Total	41	Total	35

Source: Field Survey 2008-2010

While considering the age of the head of household, it is interesting to note that in none of the villages of the study districts, the average age of the head exceeds 45 (Table 4.12) In case of Koraput district, the average age of the head is found to be 41 and 35 in respect of Rayagada district. It may be due to separation of sons after their marriage from the parents in tribal families. This indicates that, though they are lacking experience on household affairs, they are in a position to take part in the mainstream of social life.

RELIGION AND CASTE OF HHS

India is witnessing unity in diversity. The deep-rooted feelings on caste and religion have been standing as a stumbling block for national integration and developmental process. Though the study area was tribal dominated but from the scientific research angle, it was determined to know in detail about the composition of the total population from religion and caste points of view. Except Hindus and Christians, people of no other religion are found in the study area. Even some of the Christians have been converted from Hinduism since 2/3 generations.

A detailed distribution of households of Koraput district on the basis of religion is depicted in Table 4.13(a). It was found that majority of people are Hindus. Among the study villages, Aligan had the highest %age (i.e. 99.77%) of people belonging to Hindus followed by Panchada (99.47%) and Bhitaragada (99.30%). The highest percentages of people who live in Devighat (i.e. 49.02%) are Christians in comparison to other study villages. There is no presence of Christians either original or converted in Baraguda, Jhodipadar and Srirampur villages. Enquiry revealed that more than 98 per cent of Christians have been converted from Hindus since 2/3 generations. But in case of Rayagada district, as indicated in Table 4.13 (b), it reflects that except 5 households i.e. 1 (1.25 per cent) household in Gunakhal and 4 (3.67 per cent) households in Matikona, all the tribal and non-tribal households in the studied villages profess Hindu religion. The effect of Christianity is not evident in the sampled villages. Hence, there is hardly any case of conversion found on grounds of religion.

Table 4.13 (a): Distribution of HHs according to religion in Koraput district

Name of the Village	Name of the Religion				
	Hindu	Christian	Christian		Total
			Original	Converted	
Devighat	104 (50.98)	100 (49.02)	0 (0.0)	100 (100.0)	204 (100.0)
Dhaudapadar	117 (88.64)	15 (11.36)	0 (0.0)	15 (100.0)	132 (100.0)
Machput	64 (96.97)	2 (3.03)	0 (0.0)	2 (100.0)	66 (100.0)
Nighamaniguda	134 (92.41)	11 (7.59)	1 (9.09)	10 (90.91)	145 (100.0)
Bhitaragada	142 (99.30)	1 (0.70)	0 (0.0)	1 (100.0)	143 (100.0)
Baraguda	46 (100.0)	0 (0.0)	0 (0.0)	0 (0.0)	46 (100.0)
Panchada	566 (99.47)	3 (0.53)	0 (0.0)	3 (100.0)	569 (100.0)
Jhodipadar	42 (100.0)	0 (0.0)	0 (0.0)	0 (0.0)	42 (100.0)
Mukhibedei	186 (80.17)	46 (19.83)	0 (0.0)	46 (100.0)	232 (100.0)
Srirampu t	85 (100.0)	0 (0.0)	0 (0.0)	0 (0.0)	85 (100.0)
Mania	301 (90.94)	30 (9.06)	2 (6.66)	28 (93.34)	331 (100.0)
Aligan	443 (99.77)	1 (0.23)	0 (0.0)	1 (100.0)	444 (100.0)
Total	2230 (91.43)	209 (8.57)	3 (1.43)	206 (98.57)	2439 (100.0)

Note: Figures in the bracket indicate percentage to the total
Source: Field Survey 2008-2009.

Table 4.13 (b): Distribution of HHs according to religion in Rayagada district

Name of the Village	Name of the Religion				Total
	Hindu	Christian	Christian		
			Original	Converted	
Garadi	42 (100.0)	0 (0.0)	0 (0.0)	0 (0.0)	42 (100.0)
Chanchadaguda	71 (100.0)	0 (0.0)	0 (0.0)	0 (0.0)	71 (100.0)
Bhatapur	125 (100.0)	0 (0.0)	0 (0.0)	0 (0.0)	125 (100.0)
Majhihalma	36 (100.0)	0 (0.0)	0 (0.0)	0 (0.0)	36 (100.0)
Gujalpadu	74 (100.0)	0 (0.0)	0 (0.0)	0 (0.0)	74 (100.0)
K. Dengasargi	43 (100.0)	0 (0.0)	0 (0.0)	0 (0.0)	43 (100.0)
Gunakhal	79 (98.75)	1 (1.25)	1 (100.0)	0 (0.0)	80 (100.0)
Bariguda	47 (100.0)	0 (0.0)	0 (0.0)	0 (0.0)	47 (100.0)
Matikona	105 (96.33)	4 (3.67)	4 (100.0)	0 (0.0)	109 (100.0)
Laxmipur	55 (100.0)	0 (0.0)	0 (0.0)	0 (0.0)	55 (100.0)
Nathama	91 (100.0)	0 (0.0)	0 (0.0)	0 (0.0)	91 (100.0)
Tadama	143 (100.0)	0 (0.0)	0 (0.0)	0 (0.0)	143 (100.0)
Total	911 (99.45)	5 (0.55)	5 (100.0)	0 (0.0)	916 (100.0)

Note: Figures in the bracket indicate percentage to the total
Source: Field Survey 2009-2010.

Table 4.14 (a): Village-wise ethnic distribution of HHs of Koraput district

Name of the Village	Total HHs	SC	ST	OBC	General (including other religions)
Devighat	204 (100.0)	10 (4.90)	90 (44.12)	7 (3.43)	97 (47.55)
Dhaudapadar	132 (100.0)	33 (25.0)	69 (52.28)	2 (1.51)	28 (21.21)
Machput	66 (100.0)	6 (9.09)	56 (84.85)	0 (0.0)	4 (6.06)
Nighamaniguda	145 (100.0)	17 (11.72)	54 (37.24)	14 (9.66)	60 (41.38)
Bhitaragada	143 (100.0)	11 (7.69)	23 (16.09)	53 (37.06)	56 (39.16)
Baraguda	46 (100.0)	45 (97.83)	1 (2.17)	0 (0.0)	0 (0.0)
Panchada	569 (100.0)	124 (21.79)	371 (65.20)	16 (2.81)	58 (10.20)
Jhodipadar	42 (100.0)	0 (0.0)	42 (100.0)	0 (0.0)	0 (0.0)
Mukhibedei	232 (100.0)	6 (2.58)	124 (53.45)	53 (22.85)	49 (21.12)
Sriramput	85 (100.0)	0 (0.0)	45 (52.94)	8 (9.41)	32 (37.65)
Mania	331 (100.0)	72 (21.75)	78 (23.57)	48 (14.50)	133 (40.18)
Aligan	444 (100.0)	7 (1.58)	251 (56.53)	16 (3.60)	170 (38.29)
Total	2439 (100.0)	331 (13.57)	1204 (49.36)	217 (8.90)	687 (28.17)

Note: Figures in the bracket indicate percentage to the total
Source: Field Survey 2008-2009

In case of Koraput district, as revealed in Table 4.14(a), among the study villages, Panchada has the highest number of HHs i.e. 569 followed by Aligan 444. In the table, general category includes unreserved category of Hindus and other religions i.e. Christians. In the study villages, it has been found that there is presence of Hindus and Christians only. Overall data shows that out of 2439 total households, ST (49.36%) concentration is more than the other castes like SC (13.57%), OBC (8.89%) and general category (28.17%). Among the study villages, SC, OBC and general categories constitute the highest percentaage i.e. 97.83 per cent, 37.06 per cent and 47.55 per cent in Baraguda, Bhitaragada and Devighat villages respectively. In village Jhodipadar 100 per cent HHs belong to ST. But in Baraguda village only one HH is found to be ST. The maximum number of STs constitutes 84.85 per cent in Machput village followed by Panchada (65.2%) and Aligan (56.53%).

It is revealed from the study that the sub-castes under SCs are mainly Dambas, Chandalas, Panas, Harijans and Dhobis and sub-caste of STs are found mostly as Parajas, Gadabas, Kondhas, Paikas, Matias and Katia Adivasis. In Srirampt village not a single SC family is found. In other villages majority HHs belong to Dambas (SC). Among STs, Parajas (ST) have larger presence (more than 82 %) followed by Gadabas and Kondhas.

Table 4.14(b): Village-wise ethnic distribution of HHs of Rayagada district

Name of the Village	Total HHs	SC	ST	OBC	General (including other religions)
Garadi	42 (100.0)	0 (0.0)	42 (100.0)	0 (0.0)	0 (0.0)
Chanchadaguda	71 (100.0)	0 (0.0)	65 (91.55)	2 (2.82)	4 (5.63)
Bhatapur	125 (100.0)	15 (12.00)	109 (87.2)	1 (0.8)	0 (0.0)
Majhihalma	36 (100.0)	0 (0.0)	36 (100.0)	0 (0.0)	0 (0.0)
Gujalpadu	74 (100.0)	2 (2.70)	71 (95.95)	0 (0.0)	1 (1.35)
K. Dengasargi	43 (100.0)	0 (0.0)	43 (100.0)	0 (0.0)	0 (0.0)
Gunakhal	80 (100.0)	0 (0.0)	69 (86.25)	0 (0.0)	11 (13.75)
Bariguda	47 (100.0)	0 (0.0)	47 (100.0)	0 (0.0)	0 (0.0)

Name of the Village	Total HHs	SC	ST	OBC	General (including other religions)
Matikona	109 (100.0)	0 (0.0)	98 (89.91)	10 (9.17)	1 (0.92)
Laxmipur	55 (100.0)	17 (30.91)	37 (67.27)	1 (1.82)	0 (0.0)
Nathama	91 (100.0)	0 (0.0)	87 (95.60)	0 (0.0)	4 (4.40)
Tadama	143 (100.0)	48 (33.57)	94 (65.73)	1 (0.70)	0 (0.0)
Total	916 (100.0)	82 (8.95)	798 (87.12)	15 (1.64)	21 (2.29)

Note: Figures in the bracket indicate percentage to the total
Source: Field Survey 2009-2010

From the above data Table 4.14 (b) it is found that, out of 916 households of Rayagada district covered under this study, 798 households (87.12%) are scheduled tribes and 82 (8.95%) households are scheduled castes. OBC households number 15 (1.64%) whereas general caste including other religion households are 21(2.29%) only. So it is apparent that all the sampled villages predominantly consist of tribal households. Only four villages out of twelve, are having scheduled caste population. Tadama village is having largest number of scheduled caste households numbering 48 (33.57%). Seven villages viz. Garadi, Majhihalma, Gujalpadu, K. Dengasargi, Gunakhal, Bariguda and Nathama are not having any OBC households. In Matikona village, the highest number of OBC households numbering 10(9.17%) are available. Seven villages viz. Garadi, Bhatapur, Majhihalma, K. Dengasargi, Bariguda, Laxmipur and Tadama were not having any households belonging to general category including other religions. Gunakhal is having highest number of general households 11 (13.75%). Four villages namely Chanchadaguda, Majhihalma, K. Dengasargi and Bariguda were having only tribal households. It was, however, noticed that scheduled tribe dominance is more in Rayagada as compared to Koraput district as (shown in Fig. 4.3).

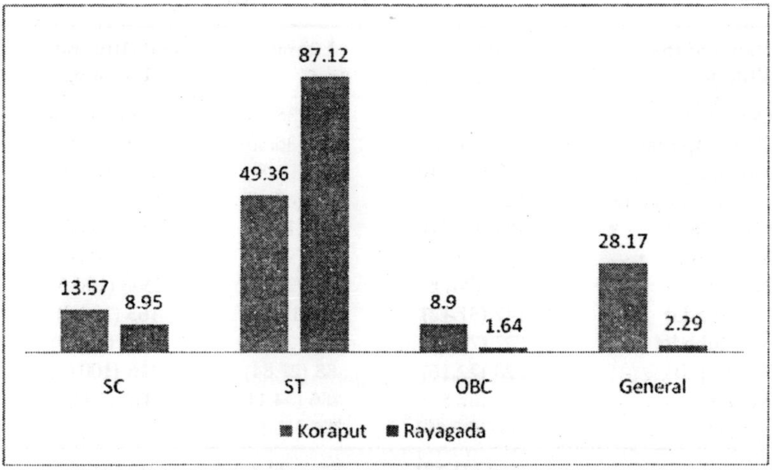

Fig. 4.3: Ethnic distribution of households in study districts

EDUCATIONAL STATUS

Education determines the way of thinking, action and behavioural pattern of any group/society. Rationality and scientific attitude is developed through education. The facilities of education are provided both by the government and non-governmental organizations. For promotion of education special emphasis is given for improvement of *dalits* and *adivasis*. Education is one of the indicators for poverty measurement. Tribal poverty is mainly due to illiteracy. Moreover, due to ignorance and lack of awareness, the tribal people have been exploited by the non-tribals. Therefore, it was considered necessary to study the educational status of the people in the villages covered under the study. The position is reflected in Table 4.15(a).

As indicated in Table 4.15 (a), on the whole, 71 per cent people are found illiterate (including children below 0-6 years) in the study villages of Koraput district. Village-wise data shows that the highest percentage of illiteracy (i.e. more than 90%) is found in Baraguda village and the lowest per centage (i.e. 37.31 %) in Nighamaniguda, which means that maximum literates are in Nighamaniguda village.

Table 4.15 (a): Literacy status in the study villages of Koraput district

Name of the Village	Illiterates	Literates	Total (Iliiterate + Literate)
Devighat	516 (61.14)	328 (38.86)	844 (100 .0)
Dhaudapadar	317 (64.70)	173 (35.30)	490 (100.0)
Machput	182 (68.94)	82 (31.06)	264 (100.0)
Nighamaniguda	247 (37.31)	415 (62.69)	662 (100.0)
Bhitaragada	484 (82.74)	101 (17.26)	585 (100.0)
Baraguda	177 (90.31)	19 (9.69)	196 (100.0)
Panchada	1752 (74.05)	614 (25.95)	2366 (100.0)
Jhodipadar	142 (87.12)	21 (12.88)	163 (100.0)
Mukhibedei	727 (70.65)	302 (29.35)	1029 (100.0)
Srirampút	228 (72.16)	88 (27.84)	316 (100.0)
Mania	939 (65.89)	486 (34.11)	1425 (100.0)
Aligan	1497 (83.49)	296 (16.51)	1793 (100.0)
Total	7208 (71.13)	2925 (28.87)	10133 (100.0)

Note: Children (0-6 years) are included while calculating illiteracy rate
Note: Figures in the bracket indicate percentage to the total
Source: Field Survey 2008-2009.

Table 4.15 (b): Literacy status in the study villages of Rayagada district

Name of the Village	Illiterates	Literates	Total (Illiterate + Literate)
Garadi	106 (61.27)	67 (38.73)	173 (100.0)
Chanchadaguda	142 (47.65)	156 (52.35)	298 (100.0)
Bhatapur	282 (55.84)	223 (44.16)	505 (100.0)
Majhihalma	96 (71.11)	39 (28.89)	135 (100.0)
Gujalpadu	186 (53.45)	162 (46.55)	348 (100.0)
K. Dengasargi	105 (55.56)	84 (44.44)	189 (100.0)
Gunakhal	166 (48.68)	175 (51.32)	341 (100.0)
Bariguda	81(55.74)	102 (44.26)	183 (100.0)
Matikona	307 (76.94)	92 (23.06)	399 (100.0)
Laxmipur	123 (82.55)	26 (17.45)	149 (100.0)
Nathama	284 (68.10)	133 (31.90)	417 (100.0)
Tadama	387 (68.99)	174 (31.01)	561 (100.0)
Total	2286 (61.82)	1412 (38.18)	3698 (100.0)

Note: Children (0-6 years) are excluded while calculating illiteracy rate
Note: Figures in the bracket indicate percentage to the total
Source: Field Survey 2009-2010.

As indicated in Table 4.15 (b), on the whole, 61.82 per cent are illiterates (including children below 0-6 years) in the study villages of Rayagada district. Village-wise data shows that highest percentage of illiteracy (i.e. more than 82%) is found in Laxmipur village and the lowest percentage (47.65) in Chancharaguda which means maximum literates are found in Bariguda village.

Table 4.15 (c): Educational level of people in study Villages of Koraput district

Name of the Village	Total Literate	Below primary	Primary	Middle	HSC	Graduate	Post-Graduate
Devighat	328	12 (3.66)	169 (51.52)	97 (29.57)	39 (11.90)	08 (2.44)	03 (0.91)
Dhauda-padar	173	03 (1.74)	67 (38.73)	58 (33.53)	42 (24.28)	01 (0.57)	02 (1.15)
Machput	82	02 (2.44)	48 (58.54)	19 (23.17)	12 (14.64)	01 (1.21)	0 (0.0)
Nighamani-guda	415	06 (1.44)	111 (26.75)	121 (29.16)	117 (28.19)	52 (12.53)	08 (1.93)
Bhitaragada	101	04 (3.96)	70 (69.31)	23 (22.77)	03 (2.97)	0 (0.0)	01 (0.99)
Baraguda	19	0 (0.0)	13 (68.43)	04 (21.05)	0 (0.0)	01 (5.26)	01 (5.26)
Panchada	614	39 (6.36)	338 (55.05)	166 (27.03)	49 (7.98)	18 (2.93)	04 (0.65)
Jhodipadar	21	01 (4.77)	15 (71.43)	05 (23.80)	0 (0.0)	0 (0.0)	0 (0.0)
Mukhibedei	302	01 (0.33)	167 (55.30)	103 (34.11)	22 (7.28)	08 (2.65)	01 (0.33)
Sriramput	88	07 (7.95)	59 (67.04)	18 (20.46)	04 (4.55)	0 (0.0)	0 (0.0)
Mania	486	08 (1.65)	256 (52.67)	167 (34.36)	39 (8.03)	16 (3.29)	0 (0.0)
Aligan	296	04 (1.35)	194 (65.55)	61 (20.61)	30 (10.13)	07 (2.36)	0 (0.0)
Total	2925	87 (2.97)	1507 (51.52)	842 (28.79)	357 (12.21)	112 (3.83)	20 (0.68)

Note: Figures in the bracket indicate percentage to the total

Note: Persons who can only read and write are also considered as literates and are grouped under below primary level

Source: Field Survey 2008-2009.

While examining the level of education of people among literates [shown in Table 4.15 (c)], it was seen that maximum people (51.52%) have studied only up to primary level, 28.8 per cent up to middle level, 12.2 per cent up to HSC, 3.83 per cent up to graduation and only 0.68 per cent have post-graduation degree.

Table 4.15 (d): Educational level of people in study Villages of Rayagada district

Name of the Village	Total Literate	Below primary	Primary	Middle	HSC	Graduate	Post-Graduate
Garadi	67	23 (34.33)	24 (35.82)	14 (20.90)	05 (7.46)	01 (1.49)	0 (0.0)
Chanchada-guda	156	58 (37.18)	58 (37.18)	9 (5.77)	9 (5.77)	22 (14.10)	0 (0.0)
Bhatapur	223	154 (69.05)	38 (17.04)	14 (2.28)	14 (6.28)	02 (0.90)	01 (0.45)
Majhihalma	39	38 (97.43)	0 (0.0)	01 (2.57)	0 (0.0)	0 (0.0)	0 (0.0)
Gujalpadu	162	42 (25.92)	84 (51.85)	17 (10.50)	11 (6.80)	05 (3.08)	03 (1.85)
K. Dengasargi	84	21 (25.0)	30 (35.72)	23 (27.38)	8 (9.52)	2 (2.38)	0 (0.0)
Gunakhal	175	36 (20.57)	34 (19.43)	66 (37.72)	35 (20.0)	03 (1.71)	01 (0.57)
Bariguda	81	30 (37.03)	24 (29.63)	17 (20.99)	10 (12.35)	0 (0.0)	0 (0.0)
Matikona	92	60 (65.22)	20 (21.74)	07 (7.60)	03 (3.26)	01 (1.09)	01 (1.09)
Laxmipur	26	20 (76.93)	06 (23.07)	0 (0.0)	0 (0.0)	0 (0.0)	0 (0.0)
Nathama	133	80 (60.16)	27 (20.30)	21 (15.79)	03 (2.25)	02 (1.50)	0 (0.0)
Tadama	174	92 (52.87)	50 (28.74)	26 (14.94)	06 (3.45)	0 (0.0)	0 (0.0)
Total	1412	654 (46.32)	395 (27.97)	215 (15.23)	104 (7.37)	38 (2.69)	06 (0.42)

Note: Figures in the bracket indicate percentage to the total

Note: Persons who can only read and write are also considered as literates and are grouped under below primary level

Source: Field Survey 2009-2010.

While examining the education level of people among the literates in Rayagada district, it is seen that 27.97 per cent are educated up to primary level, 15.23 per cent up to middle level, 7.37 per cent up to HSC, 2.69 per cent up to graduation and only 0.42 per cent have post- graduate qualification [Table 4.15 (d)] Among the study villages, the highest illiteracy (82.55) is found in Laxmipur village and maximum literates (52.35) are reported from Chanchuraguda village.

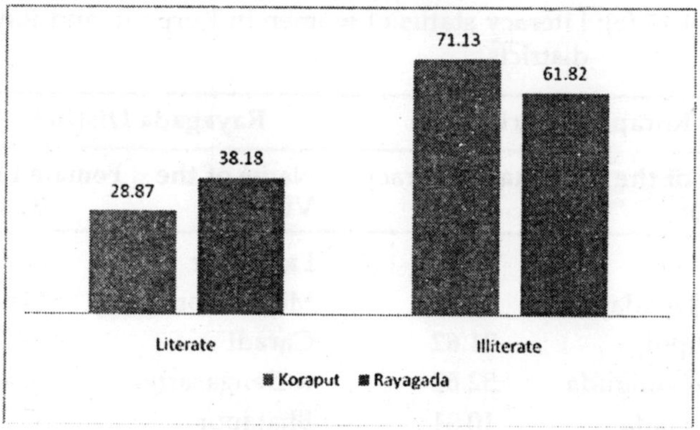

Fig.4.4: Literacy status in study districts

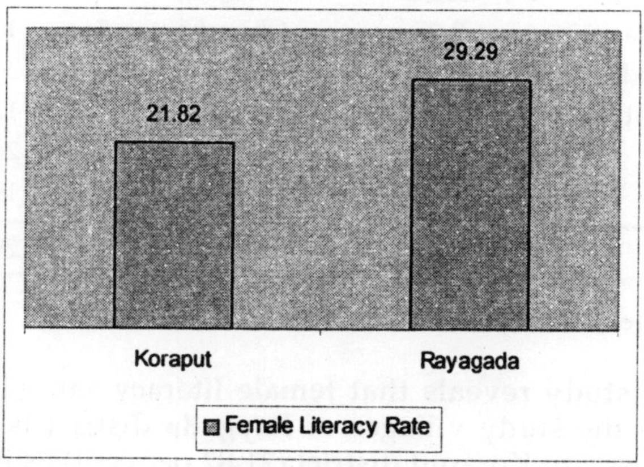

Fig.4.5: Female literacy status in study districts

LITERACY STATUS OF WOMEN

It is said that if the 'mother is educated, the total family is educated'. It signifies the importance of female education. Education is the centre of all developmental efforts. Therefore, as an indicator of development, this study attempted to make an analysis of female literacy rate in the study villages.

Table 4.15 (e): Literacy status of women in Korpaut and Rayagada districts

Koraput District		Rayagada District	
Name of the Village	**Female Literacy Rate**	**Name of the Village**	**Female Literacy Rate**
Devighat	31.82	Laxmipur	14.70
Dhaudapadar	28.35	Majhihaloma	29.48
Machhput	21.62	Garadi	25.77
Nighamaniguda	52.63	K.Dengasarigi	36.45
Bhitaragada	10.31	Bhatapur	37.72
Baraguda	11.7	Gujalpadu	35.56
Panchada	18.98	Bariguda	38.70
Jhodipadar	8.97	Chancharaguda	40.62
Mukhibedei	23.10	Gunakhal	36.06
Sriramput	20.24	Tadama	17.44
Mania	25.20	Nathama	24.13
Aligan	9.33	Matikona	19.35
Total	21.82	Total	29.29

Source: Field Survey 2008-2010

The study reveals that female literacy rate (29.29 per cent) in the study viilages of Raygada district is more in comparison to Koraput district (21.82 per cent) as indicated in Table 4.15(e). Nighamaniguda village of Koraput district witnesses the highest female literacy (52.63) per cent while this type of situation is found in Chancharaguda village (40.62) per cent of Raygada district. Likewise, the lowest literacy rate (9.33 per cent) has been reported in Aligan of

Koraput district while it is (14.70 per cent) in Laxmipur village of Raygada district. Thus literacy status is found to be better in Rayagada district than in Koraput district (shown in Figure 4.4 and 4.5)

It can be inferred that the people with a relatively poor economic background did not realize the immediate relevance of female education and preferred to send their children for labour work to earn livelihood for the families. As a result of their unenthusiastic attitude towards education in the given context, their socio-economic background, the rate of female literacy is quite low in both the districts. It is observed that, specifically the tribal girls do not find any time for schooling to come out of their household works. Parents also are not very particular about their daughters' education. In fact, the present availability of minimum educational facilities in the villages is unable to attract students of the tribal groups to its structural and functional areas due to several socio-economic and technical factors. On the other hand, the low literacy rate and low performance in higher studies put questions on the availability of educational facilities in the villages and the extent of opportunities provided to achieve education. It has been observed that due to lack of regular flow of income, the general level of education is low. Although the educational institutions have spatial effect on the availability and opportunities to rural set-ups, the economic condition of households is still the key determining factor for acquiring required opportunities.

OCCUPATIONAL PATTERN

The present study categorized the total occupational structure into different occupations. They are agriculture, agricultural labour, collection of MFP, horticulture, crafts and cottage industries, government services, private services, wage labour, animal husbandry and business etc. But for analysis purpose, some of the identical categories are combined into one group. The status of occupational pattern of study villages of Koraput and Rayagada district is shown in Table 4.16.

Table 4.16: Distribution of workforce

Category of occupation	Koraput District		Rayagada District	
	No. of persons in work-force	No. of STs in work-force	No. of persons in work-force	No. of STs in work-force
Agriculture & agriculture-labour	1402 (25.68)	761 (26.09)	841 (48.33)	811 (53.11)
Wage labour	3409 (62.46)	1950 (66.85)	684 (39.31)	556 (36.41)
Service holder (Pvt., Govt., Pensioners)	240 (4.40)	57 (1.95)	94 (5.40)	77 (5.04)
Other workers	407 (7.46)	149 (5.11)	121 (6.96)	83 (5.44)
Total number of labour-engagements	5458 (100.0)	2917 (100.0)	1740 (100.0)	1527 (100.0)
Proportion of workforce-to total population	53.86	242.27	42.23	42.18
Total Population	10133	1204	4120	3620

Note: Agriculture, agriculture labour, horticulture and animal husbandry named as agriculture and agriculture labour
Note: Government service, private service and pensioners as service holders
Note: Wage labourer exclusively
Note: Rest of the occupations are as in case of other workers
Source: Field Survey 2008-2010

In Koraput district, out of 10133 total population, 53.8 per cent are in workforce. It has been noticed that some of the labourers are engaged in more than one occupation. Maximum numbers of labourers (62.46%) are coming under wage labour category and only 25.68 per cent labourers are in the category of agriculture and agricultural labour. Very less number (4.4%) of persons are engaged as labourers and the rest 7.46 per cent are engaged in other works. While considering the occupational pattern of STs, the same trend is noticed (Table 4.16) i.e. maximum (66.85%) people are engaged in wage labour. Dependence on agriculture and as agriculture labourer is only 26 per cent. Only about 2 per cent are service holders and 5 per cent of total are engaged in other occupations. This data contradicts the general view that majority tribal people depend on agriculture. May be this is due to landlessness or their not being able to cultivate the available land as a result of which they have been diverted to other occupations mainly towards wage earning.

Regarding distribution of workforce in Rayagada district, it was seen that out of 4120 total population of study villages, 42.23 per cent are in workforce. Persons engaged in agriculture and as agricultural labourer are found to be the highest in number i.e. 841 (48.33%) followed by 684 (39.31%) daily wage labour category. Only 94 (5.4%) are in services and the rest 121 (6.96%) engaged in other occupations. The occupational pattern of scheduled tribes is also having the same trend. As shown in Table 4.16, out of total 1527 workforce (42.18%) of total population, the highest number of persons i.e. 811 (53.11%) are engaged in agriculture and as agricultural labour followed by 556 (36.4%) as daily wage labour. Very less number of persons i.e., 77 (5.04%) are in services and the rest 83 (5.44%) in other occupations. This indicates the predominance of households mainly dependent on agriculture in the capacity of either owners or tenants of land or as agricultural labourer. A comparison of two districts indicates that the workers are engaged more in agriculture and agricultural labour in Rayagada district whereas proportionately more are in daily wage earning in Koraput district (shown in Fig.4.6 and 4.7). Similar is the case of women workforce (shown in fig-4.8 and 4.9).

Due to weak agricultural economy coupled with low level farm technology, low productivity has adversely affected their lives. The income drawn from agriculture is very low due to low per capita land holding and area covered, area sown as few irrigation facilities are available for them. Although the percentage of households engaged in government and private employment is there but its impact in alleviating the tribal economy is not visible significantly. Even though a few persons are engaged in regular employment both in public and private sectors, still they prefer agriculture.

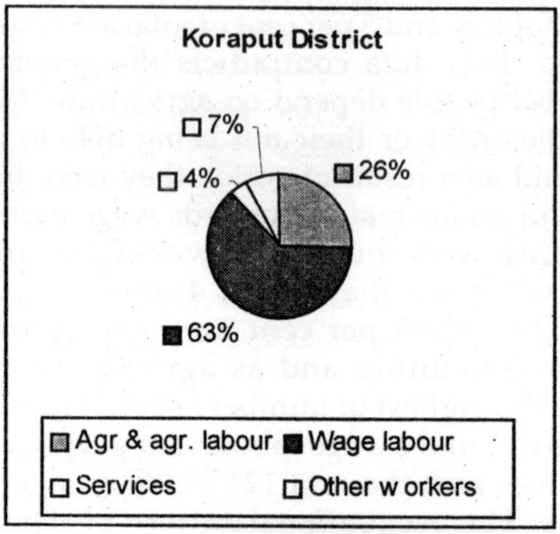

Figure 4.6: Distribution of workforce in Koraput district

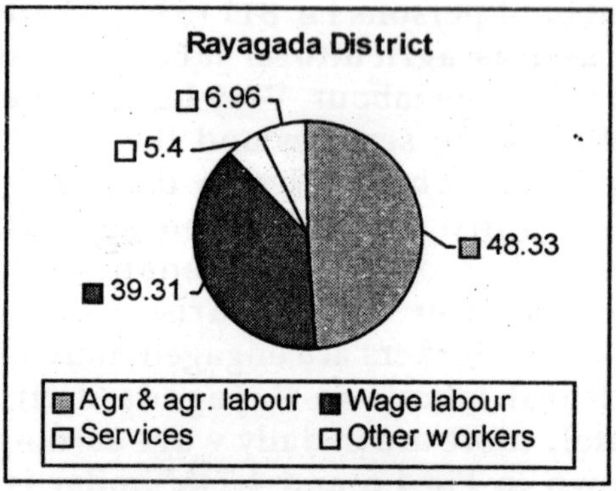

Figure 4.7: Distribution of workforce in Rayagada district

FEMALE WORKFORCE:

Contribution of women to the household economy is immense especially in tribal areas. Tribal women struggle for their livelihood on one side and engage themselves in activities of household chores on the other. Mostly, male members are found highly alcoholic in tribal areas. As a result, women members are toiling day and night. Occupational structure of women in Koraput and Rayagada district is reflected in Table 4.17. Out of 1883 (36.36%) total female workforce in Koraput district, maximum women (77.8%) are engaged as daily wage labour. About 7.6 per cent are engaged in agriculture and agricultural labour, 3.14 per cent in services and the rest 11.47 per cent in other works. Similarly, in Rayagada district, out of 608 (27.22%), proportionately more women (48.03%) are engaged as daily wage labour than women engaged in agriculture and agriculture labour (46.38%) and (2.30%) in services and the rest (3.29%) in other works.

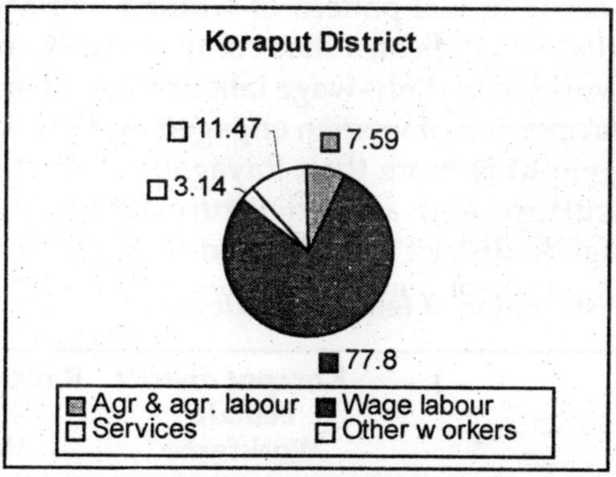

Figure 4.8: Distribution of female workforce in Koraput district

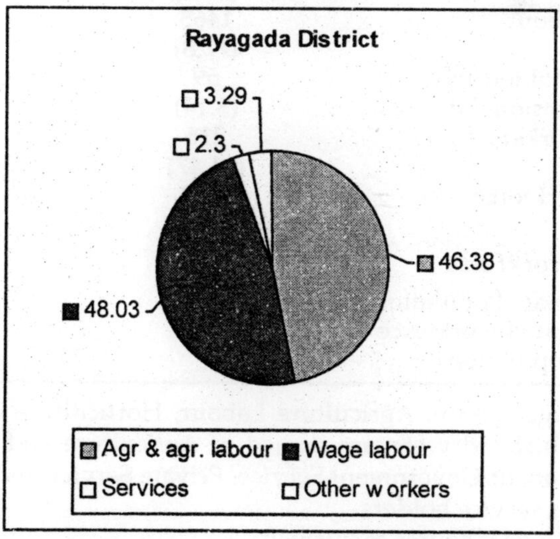

Figure 4.9: Distribution of female workforce in Rayagada district

The study shows that in both the districts, occupational pattern is slightly different from each other:

(i) Occupational pattern of workforce (both male and female), in respect of majority of persons in Koraput district is wage-labour based whereas in Rayagada district more seem to be engaged in agriculture and as agriculture labours compared to other occupations.

(ii) The occupational pattern of women workforce in both the districts indicates that comparatively more people are working as daily-wage labour than other works. But the proportion of women engaged as daily wage-labour in Koraput is more than Rayagada district. Women in agriculture and as agriculture labour are more in Rayagada district as compared to Koraput district

Table 4.17: Distribution of female workforce

Category of Occupation	Koraput district Female Workforce	Rayagada district Female Workforce
Agriculture & agriculture-labour	143 (7.59)	282 (46.38)
Wage Labour	1465 (77.8)	292 (48.03)
Service Holder (Pvt., Govt., Pensioners)	59 (3.14)	14 (2.30)
Other Workers	216 (11.47)	20 (3.29)
Total Workforce	1883 (100.0)	608 (100.0)
Female workforce		
Total Female Population	5178	2233
Proportion of workforce to-female population	36.36	27.22

Notes: Agriculture, Agriculture Labour, Horticulture and Animal Husbandry etc. are named as Agriculture and Agriculture Labour Government Service, Private Service and Pensioners as service holders.
Wage labourer exclusively.
Rest of the occupation as Other Workers

Source: Field Survey 2008-2010.

INCOME AND CONSUMPTION PATTERN

Income of the household symbolizes both social and economic prosperity of the people. An effort was made to know the income level of the people in the study villages. The present survey made every possible effort to collect reliable information about the annual income of tribal households in the study villages of the select districts. The respondents of the households could not provide exact information about their monetary income. The researchers tried to know about their produce on the basis of this information available in regard to the annual income of the households as estimated. Table 4.18 (a) and (b) indicate the annual income of the study villages of Koraput and Rayagada districts respectively. Income of HHs of majority of villages of Koraput district (50.4%) comes under the range of Rs.10,000 to Rs. 20,000 per annum. Among all the villages, the highest number of people (37.93%) of Nighamaniguda are earning a sizable income which varies from Rs. 40,000 to Rs.50,000 followed by village Mania (34.74%) in the range of Rs 20,000 to Rs. 30,000 annually. This indicates that people of Nighamaniguda are better than other villages. The economic position of Bhitaragada is the worst among all the villages. The pattern of low income may be attributed to gradual decline of dependence on agriculture. As because the tribals have either very less quantity of land in possession or are not in a position to cultivate such lands in their possession or they may have no diversified employment opportunities.

Table 4.18 (a): Annual income of the study villages of Koraput district

Name of the Village	No. of households having annual income (in Rs.)					
	1-10,000	10,000-20,000	20,000-30,000	30,000-40,000	40,000-above	Total
Devighat	0	98	66	22	18	204
	(0.0)	(48.04)	(32.35)	(10.78)	(8.83)	(100.0)
Dhaudapadar	0	47	52	14	19	132
	(0.0)	(35.60)	(39.40)	(10.60)	(14.40)	(100.0)
Machput	0	49	14	2	1	66
	(0.0)	(74.24)	(21.22)	(3.03)	(1.51)	(100.0)

Name of the Village	No. of households having annual income (in Rs.)					
	1-10,000	10,000-20,000	20,000-30,000	30,000-40,000	40,000-above	Total
Nighamaniguda	0 (0.0)	52 (35.86)	25 (17.24)	13 (8.97)	55 (37.93)	145 (100.0)
Bhitaragada	0 (0.0)	108 (75.53)	27 (18.88)	7 (4.89)	1 (0.70)	143 (100.0)
Baraguda	0 (0.0)	44 (95.65)	2 (4.35)	0 (0.0)	0 (0.0)	46 (100.0)
Panchada	56 (9.84)	243 (42.71)	162 (28.47)	73 (12.83)	35 (6.15)	569 (100.0)
Jhodipadar	0 (0.0)	30 (71.43)	11 (26.19)	1 (2.38)	0 (0.0)	42 (100.0)
Mukhibedei	19 (8.19)	118 (50.86)	74 (31.90)	10 (4.31)	11 (4.74)	232 (100.0)
Sriramput	16 (18.82)	53 (62.35)	15 (17.65)	0 (0.0)	1 (1.18)	85 (100.0)
Mania	24 (7.25)	101 (30.51)	115 (34.74)	60 (18.13)	31 (9.37)	331 (100.0)
Aligan	88 (19.82)	287 (64.64)	57 (12.84)	6 (1.35)	6 (1.35)	444 (100.0)
Total	203 (8.32)	1230 (50.43)	620 (25.42)	208 (8.54)	178 (7.29)	2439 (100.0)

Note: Figures in the bracket indicate percentage to the total
Source: Field Survey 2008- 2009.

Table 4.18 (b): Annual income of the study villages of Rayagada district

Name of the Village	No. of households having annual income (in Rs.)					
	1-10,000	10,000-20,000	20,000-30,000	30,000-40,000	40,000-above	Total
Garadi	4 (9.52)	11 (26.20)	19 (45.24)	4 (9.52)	4 (9.52)	42 (100.0)
Chanchadaguda	7 (9.86)	13 (18.31)	24 (33.80)	14 (19.72)	13 (18.31)	71 (100.0)

Name of the Village	No. of households having annual income (in Rs.)					
	1-10,000	10,000-20,000	20,000-30,000	30,000-40,000	40,000-above	Total
Bhatapur	35 (28.0)	67 (53.6)	22 (17.60)	1 (0.8)	0 (0.0)	125 (100.0)
Majhihalma	7 (19.44)	18 (50.00)	9 (25.00)	2 (5.56)	0 (0.0)	36 (100.0)
Gujalpadu	0 (0.0)	33 (44.59)	31 (41.89)	6 (8.11)	4 (5.41)	74 (100.0)
K. Dengasargi	1 (2.32)	27 (62.80)	7 (16.28)	7 (16.28)	1 (2.32)	43 (100.0)
Gunakhal	1 (1.25)	36 (45.0)	19 (23.75)	12 (15.0)	12 (15.0)	80 (100.0)
Bariguda	3 (6.38)	23 (48.94)	8 (17.02)	8 (17.02)	5 (10.64)	47 (100.0)
Matikona	6 (5.50)	38 (34.87)	30 (27.52)	22 (20.18)	13 (11.93)	109 (100.0)
Laxmipur	12 (21.82)	19 (34.55)	14 (25.46)	6 (10.90)	4 (7.27)	55 (100.0)
Nathama	5 (5.49)	7 (7.69)	13 (14.29)	36 (39.56)	30 (32.97)	91 (100.0)
Tadama	3 (2.10)	70 (48.95)	50 (34.96)	13 (9.09)	7 (4.90)	143 (100.0)
Total	84 (9.17)	362 (39.52)	246 (26.85)	131 (14.31)	93 (10.15)	916 (100.00)

Note: Figures in the bracket indicate percentage to the total
Source: Field Survey 2009-2010

It is found from Table 4.18 (b) that the highest percentage (39.52%) of the households are in the range of Rs. 10,000 to Rs. 20,000 income group followed by (26.85%) in the range of Rs. 20,000 to Rs. 30,000 income group. Fourteen per cent of the households are in Rs. 30,000 to Rs. 40,000 income group, 10 per cent of the households are in the Rs. 40,001 above income group. Among the study villages, the economic position of Nathama is better than other villages as about 40 per cent of total HHs earn within Rs. 30000 to Rs. 40000 annually. Similarly, economic position of K. Dengasarigi is the worst among all the study villages as more than 60 per cent HHs come under Rs. 10000 to Rs. 20000 income range. Like income estimation, expenses of each head were calculated on the basis of number of units consumed. While enquiring about the average expenditure of HHs of study villages of Koraput district, it was found that HHs spent more on food i.e. Rs. 12,690 annually than other items. They incur very less amount on maintenance of house which is only Rs. 729 per annum. The average expenditure of ST HH on food is about Rs. 12,276 annually compared to other items of expenditure. They incur very less amount on intoxicants (Rs. 687 per annum) like *bidi*, leaf for smoking (locally made), tobacco etc. In Rayagada district, major portion of income is spent on food and very less amount is incurred on intoxicants. Since the villagers could not give exact information about income and expenditure position in terms of money, the data may not be accurate as mentioned. The researchers tried to extract more or less the approximate expenditure position as far as practicable.

KEY FINDINGS

1. In Koraput district, total of 2439 households covering a population of 10133 comprising of 48.9 per cent and female population of 5178 (51.1%) were surveyed for the purpose. Among the 12 villages of Koraput district, Jhodipadar has the highest percentage of male population (52.15%) whereas female population is found to maximum in Machput village (43.94%). In case of Rayagada district, a total of 916 households having 4120 population comprising of 1923 (46.63%) are male and 2197 (53.33%) females. It also reveals that, except two villages viz. Bariguda and Laxmipur, all other studied villages are female dominated.
2. Average size of the family in the study villages of Koraput is 4.1 and in Rayagada 4.5 respectively.
3. In Koraput district, the average age of the head of household is 41 but it is 35 in Rayagada district. The average size of family is almost 4 in both the districts.

Both the age of the head of household and size of the family is low because sons are separated from their parents immediately after their marriage.

4. In Koraput district, 91.43 per cent people are Hindus. Village-wise data shows that Aligan has the highest percentage (99.77%) of people belonging to Hindus followed by Panchada (99.47%) and Bhitaragada (99.30%). The highest percentage of Christian people is found in Devighat (49.02%) village. In Rayagada district, except 5 households only one (1.25 %) household in Gunakhal and 4 (3.67%) households in Matikona, all the tribal and non-tribal households in the studied villages profess Hindu religion. The effect of Christianity is not evident in the sampled villages.
5. Overall data of Koraput district shows that out of 2439 total households, ST (49.36%) concentration is more than other castes like SC (13.57%), OBC (8.90%) and general category (28.17%). Out of 916 households in Rayagada district, 798 (87.12%) households are scheduled tribes and 82 (8.95%) households are scheduled caste. OBC households constitute 15 (1.64%) whereas 21(2.29%) households belong to general caste population including other religion. Scheduled tribe dominance is more in Rayagada as compared to Koraput district.
6. In Koraput district, 71 per cent of people are illiterate including children below 0-6 years. Among the literates, about 51 per cent are educated up to primary level, 3.83 per cent only up to graduate level and 0.68 per cent have studied up to post-graduation level. In the study villages, the highest percentage (63%) of people in Nighamaniguda are literate and the lowest number of literates (10%) is found in Baraguda village. In Rayagada district, 1412 (38.18%) are literates and the rest 2286 (61.82%) are illiterates. Among the study villages, the highest illiteracy (82.55) is found in Laxmipur village and maximum literates (52.35) are reported in Chanchuraguda village. Among the literates, 27.97 per cent are up to primary level, 2.69 per cent up to graduation and only 0.42 per cent have post-graduate qualification.

7. While comparing the literacy status of both the districts, a significant gap is noticed between the two districts. This might be due to difference in literacy calculation method (mentioned in the note under 4.15(b) &(c)). In Koraput district children between 0-6 years of age are included whereas in case Rayagada district, children between 0-6 years are excluded because of non-availability of authentic data.
8. The study reveals that female literacy rate (29.29%) in the study villages of Raygada district is more in comparison to Koraput district (21.82%). Nighamaniguda witnesses the highest female literacy (52.63%) in Koraput district while this type of case is found in Chancharaguda (40.62%) of Raygada district. Likewise, the lowest literacy rate (9.33%) has been reported in Aligan of Koraput district but it is (14.70%) in Laxmipuru village of Raygada district. The study revealed that literacy status is better in Rayagada district than in Koraput district.
9. In the study area of Koraput district, 53.86 per cent are in workforce. It has been noticed that some of the labourers are engaged in more than one occupation. Maximum number of labourers (62.46%) are coming under wage-earning labour category and only 25.68 per cent labourers are in the category of agriculture and agricultural labourer. About service holders, 4.4 per cent of persons are found and the rest 7.46 per cent are engaged in other works. In occupational pattern of STs, the same trend is noticed i.e. maximum (66.85%) are engaged in wage labour. Dependence on agriculture and as agriculture labour is only 26 per cent. About 2 per cent are service holders and 5 per cent of total workforce are engaged in other occupations. The total workforce in the study area of Rayagada district is 42.23 per cent. Persons engaged in agriculture and as agricultural labours are respectively found to be 841 (48.33%) followed by 684 (39.31%) as daily wage labour. Only 94 (5.4%) are in services and the rest 121 (6.96%) are engaged in other occupations. Out of total 1527 workforce persons, 811 (53.11%) are engaged in

agriculture and agricultural labour followed by 556 (36.4%) as daily-wage labour. Only 77 (5.04%) persons are in services and the rest 83 (5.45%) in other occupations.

10. In case of occupational pattern of workforce (both male and female), majority of persons in Koraput district are engaged in wage labour whereas more are in agriculture and agriculture labour in Rayagada district as compared to other occupations.
11. While comparing the occupational pattern of women workforce in both the districts, it was noticed that comparatively more persons are working as daily wage-labour than other works. But the proportion of women engaged in daily wage-labour in Koraput is more than Rayagada district. Women working in agriculture and as agriculture labour are found to be more in Rayagada district than Koraput district.
12. Income of HHs of majority of villages of Koraput district (50.4%) stands in the range Rs. 10,000 to Rs. 20,000. Among all the villages, the highest percentage of people (37.93%) of Nighamaniguda are earning a sizable income ranging from Rs. 40,000 to Rs. 50,000 followed by village Mania with 34.74 per cent in the range of Rs. 20,000 to Rs. 30,000 annually. This indicates that people of Nighamaniguda are better than other villages in terms of income. Similarly, the economic position of Bhitaragada is the worst among all the villages in Koraput district. The highest per centage (39.52%) of the households of Rayagada district are in the income range of Rs. 10,000 to Rs. 20,000, followed by 26.85 per cent in the range of Rs. 20,000 to Rs. 30,000 income group and 14 per cent of the households are within Rs. 30,000 to Rs. 40,000 income group but 10.13 per cent of the households are in the Rs. 40,001 above income group. Among the study villages, the economic position of Nathama is better than other villages as highest percentage (32.98%) of households come within the income group and 39.56 per cent come in the range Rs. 20,000 to Rs. 30,000 and Rs. 40,001 and above respectively. The economic

position of K. Dengasargi village is the worst among all study villages in Rayagada district. On the whole between the two districts, the economic position in Rayagada district seems to be better than Koraput district.

13. The average expenditure of HHs of study villages of Koraput district, on food i.e. Rs. 12,690 annually is more than on other items. They incur very less amount on maintenance of house, which is only Rs. 729 per annum. In case of Rayagada district, major portion of income is spent on food i.e. Rs. 12,880 annually and lowest amount i.e. Rs. 548 is incurred on intoxicants.

5

Land Holdings and Land-Use Pattern

This chapter analyses the nature of ownership of land, landholding and land use pattern of different categories of land, average size of landholding and operational holdings in the study villages of Koraput and Rayagada districts.

One of the major problems for the vast majority of the rural population in India is the inadequate or almost non-existent land holdings. Rural poverty in India is attributed largely to the absence of access to land. Secured access rights to land is also an imperative for food security. Land holding is an outstanding and inseparable economic feature of the tribal society. Size of land holdings is one of the factors which determine the degree of poverty. So, the type and size of land holding plays a vital role in an agricultural economy and more so in tribal economy. In the study area of Koraput and Rayagada districts, the agricultural land is categorized broadly into three types: (i) Dangara, (ii) Paddy, and (iii) Atala. Each category is again divided into three types i.e. I, II, III on the basis the degree of productivity. In Paddy lands, paddy is cultivated exclusively. Landowners cultivate ragi, turmeric, bazra etc. in Dangara type land and maize, ragi etc. in Atala land. The present chapter tries to examine the status of land holding pattern in both the select districts. Ownership of land holding is the major issue among all other issues related to land. It not only provides economic security but also social status in society.

NATURE OF OWNERSHIP OF LAND

The last survey and settlement of land was done during 1956-64 in Koraput and Rayagada districts. Thereafter, no survey and settlement operation has taken place. In consequence, RoRs of land have not been updated although these could be done through mutation when settlement operations are not in force. But due to lack of awareness and increase in the number of legal heirs, RoRs of original landowners have not been transferred to their respective children and grandchildren in majority cases. The grandchildren possess those lands by inheritance. Hence in the study area, it was difficult on the part of the researchers to find out the exact number of landowners, because the persons who possess land by inheritance do not have RoRs recorded in their names. Due to the above factors, the nature of ownership over land could be divided into three categories viz. by (i) "own possession", (ii) "possession by inheritance" and (iii) "no possession". "Own possession" includes *Patta* holder of *Raiyat* land and lease land. In some cases, the landholders are having encroached lands also. "Possession by inheritance" implies that although persons have no record of rights but they are enjoying land by inheritance and/or in occupation through verbal/mutual transfer. But there could be some cases of possession of encroached lands. Table 5.1 (a) indicates the status of land in the villages in Koraput district covered under the study. "No possession" refers to those HHs who have neither RoRs nor enjoying the land due to inheritance and/or in the kind of encroached land.

Table 5.1 (a): Status of ownership of land in Koraput district

Name of the village	Status of ownership			Total
	Own possession	Possession through inheritance	No possession	
Devighat	41 (20.10)	18 (8.82)	145 (71.08)	204 (100.0)
Dhaudapadar	24 (18.18)	1 (0.76)	107 (81.06)	132 (100.0)

Name of the village	Status of ownership			Total
	Own possession	Possession through inheritance	No possession	
Machhput	28 (42.43)	6 (9.09)	32 (48.48)	66 (100.0)
Nighamaniguda	28 (19.32)	14 (9.65)	103 (71.03)	145 (100.0)
Bhitaragada	77 (53.85)	25 (17.48)	41 (28.67)	143 (100.0)
Baraguda	9 (19.56)	11 (23.92)	26 (56.52)	46 (100.0)
Panchada	183 (32.16)	123 (21.62)	263 (46.22)	569 (100.0)
Jhodipadar	28 (66.67)	1 (2.38)	13 (30.95)	42 (100.0)
Mukhibedei	23 (9.91)	130 (56.03)	79 (34.06)	232 (100.0)
Sriramput	17 (20.0)	36 (42.36)	32 (37.64)	85 (100.0)
Mania	36 (10.88)	174 (52.57)	121 (36.55)	331 (100.0)
Aligan	157 (35.36)	116 (26.13)	171 (38.51)	444 (100.0)
Total	651 (26.69)	655 (26.86)	1133 (46.45)	2439 (100.0)

Note: Figures in the bracket indicate percentage to the total
Source: Field Survey 2008-2009.

Table 5.1 (a) reveals the data on the nature of ownership in Koraput district. It has been found that majority of the villagers (81.06%) in Dhaudapadar village have no land at all. Among the study villages, Jhodipadar has the maximum number of HHs having *patta* land (66.67%) followed by Bhitaragada (53.85%). Very less number of people in Mukhibedei (9.91%) are in possession of land. But the maximum number of villagers (56.03%) enjoy land by inheritance. In this case, it was very difficult to assess about the exact position of landless HHs. Overall, 651 HHs constituting 26.69 per cent HHs of study villages possess land of their own, 655 HHs constituting 26.16 per cent by inheritance and 1133 HHs constituting 46.45 per cent are landless. This indicates that more than 50 per cent are having land in either way. As per the classification, 651 are recorded landholders, 655 HHs have acquired land by inheritance as shown in Figure 5.1 (a).

Table 5.1(b) indicates the position of land ownership in Rayagada district.

Table 5.1 (b): Status of ownership of land in Rayagada district

Name of the village	Status of ownership			Total
	Own possession	Possession through inheritance	No possession	
Garadi	17 (40.48)	22 (52.38)	3 (7.14)	42 (100.0)
Chanchadaguda	22 (30.98)	16 (22.54)	33 (46.48)	71 (100.0)
Bhatapur	43 (34.40)	56 (44.80)	26 (20.80)	125 (100.0)
Majhihalma	1 (2.78)	25 (69.44)	10 (27.78)	36 (100.0)
Gujalpadu	28 (37.84)	41 (55.40)	5 (6.76)	74 (100.0)
K. Dengasargi	16 (37.21)	24 (55.81)	3 (6.98)	43 (100.0)
Gunakhal	36 (45.00)	29 (36.25)	15 (18.75)	80 (100.0)
Bariguda	17 (36.17)	13 (27.66)	17 (36.17)	47 (100.0)
Matikona	15 (13.76)	60 (55.05)	34 (31.19)	109 (100.0)
Laxmipur	3 (5.45)	18 (32.73)	34 (61.82)	55 (100.0)
Nathama	14 (15.38)	30 (32.97)	47 (51.65)	91 (100.0)
Tadama	11 (7.69)	94 (65.74)	38 (26.57)	143 (100.0)
Total	223 (24.35)	428 (46.72)	265 (28.93)	916 (100.0)

Note: Figures in the bracket indicate percentage to the total
Source: Field Survey 2009-2010.

The nature of ownership of land in Rayagada district as indicated in Table 5.1 (b), shows that, 223 (24.35 per cent) households are having land in their own possession, 428 households constituting 46.72 per cent possess land through inheritance and 265 (28.93%) have no possession at all [shown in Fig.5.1 (b)]. They are considered as landless. In Gunakhal, highest number 36 (45 per cent) of households are having possession over land whereas in village Tadama 94 (65.74 percent) and in Laxmipur 34 (61.82 per cent) households are having highest percentage through inheritance and under landless categories respectively. In the present study, those of the HHs who even have acquired land by encroachment were taken under the category of landholders. Some *patta* holders also occupy encroached land apart from *Ryot* lands. Similarly, persons enjoying land by inheritance also possess encroached land for cultivation. Hence total number of landholders is calculated by adding number of landholders both in own possession and those having either lease or encroached lands. It is, therefore, clear that addition of the number of encroachment landholders, lease landholders, *patta* holders and landless will also be equal to total households. In the present study, the number of landholders of Koraput district is found to be 864 [mentioned in Table 5.8(a)], out of which, 651 are having their own *patta* lands and the rest 213 households presumably possessing lease/encroached lands exclusively. Similarly, in case of Rayagada district, out of 916 households, 223 households presumably possessing own *patta* lands and 300 having lease/encroached lands exclusively.

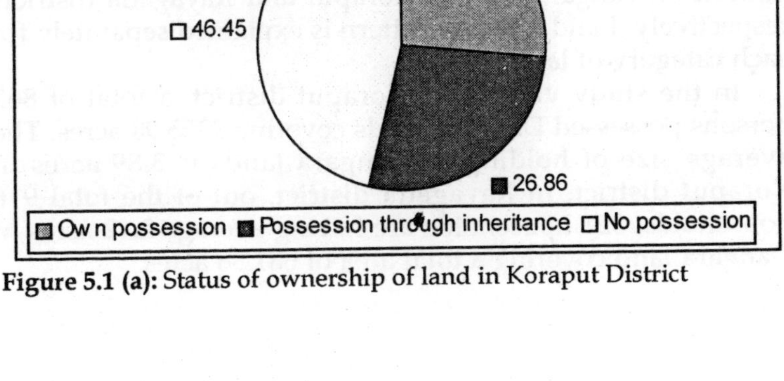

Figure 5.1 (a): Status of ownership of land in Koraput District

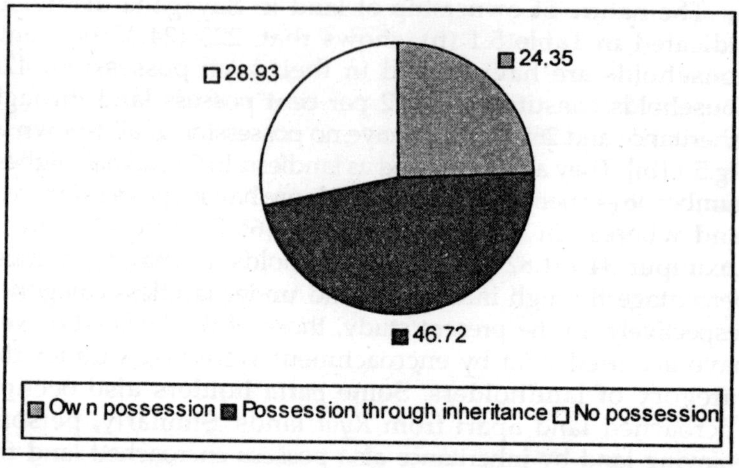

Figure 5.1 (b): Status of ownership of land in Rayagada district

The households in the study area appear to be normally owning land in three ways viz. (i) own *patta* land, (ii) lease land, and (iii) encroached land. A person may possess either one or a combination of any two or three of the above categories of lands. Land holding of all categories of land (Dangara, Paddy and Atala) mentioned in the following tables is the combination of all three types of possessions i.e, *patta* land, lease land and encroached land. The position of land holdings of lease and encroached land is explained separately.

LAND HOLDING PATTERN OF DANGARA LAND

Tables 5.2 (a) and 5.2 (b) present the data on landholding pattern of Dangara land in Koraput and Rayagada districts respectively. Land holding pattern is explained separately for each category of land.

In the study villages of Koraput district, a total of 862 persons possessed Dangara lands covering 3355.98 acres. The average size of holding of Dangaıa lands is 3.89 acres in Koraput district. In Rayagada district, out of the total 916 households, 321 households are having average 2.12 acres of Dangara land covering a total area of 681.36 acres.

Table 5.2 (a): Dangara type land in Koraput district (in acres)

Village name	Dangara-I (D-I) land owned	Dangara-II (D-II) land owned	Dangara -III (D-III) land owned	Total Dangara land owned (I+II+III)
Devighat				
Avg. size of land holding	1.72	1.31	1.14	**1.43**
No. of persons possessed	8	8	5	**21**
Extent of area covered	13.77	10.5	5.72	**29.99**
Dhaudapadar				
Avg. size of land holding	2.99	2.43	5.92	**4.13**
No. of persons possessed	7	7	11	**25**
Extent of area covered	20.99	17	65.18	**103.17**
Machhput				
Avg. size of land holding	5.96	3.85	3.85	**4.55**
No. of persons possessed	12	12	12	**36**
Extent of area covered	71.59	46.25	46.17	**164.01**
Nigamaniguda				
Avg. size of land holding	4.31	3.67	3.19	**3.74**
No. of persons possessed	14	13	13	**40**
Extent of area covered	60.37	47.74	41.45	**149.56**
Bhitaragada				
Avg. size of land holding	1.59	0.63	1.64	**1.44**
No. of persons possessed	52	14	20	**86**
Extent of area covered	82.73	8.81	32.73	**124.27**
Baraguda				
Avg. size of land holding	4.9	15.45	4.55	**7.05**
No. of persons possessed	16	5	3	**24**
Extent of area covered	78.4	77.25	13.66	**169.31**
Panchada				
Avg. size of land holding	1.87	3.512	4.08	**3.20**
No. of persons possessed	110	75	107	**292**
Extent of area covered	206.34	263.4	437.0	**906.74**
Jhodipadar				
Avg. size of land holding	0.68	1.57	1.54	**1.00**
No. of persons possessed	14	6	2	**22**
Extent of area covered	9.6	9.4	3.09	**22.09**
Mukhibedai				
Avg. size of land holding	4.50	4.05	3.06	**3.63**

Village name	Dangara-I (D-I) land owned	Dangara-II (D-II) land owned	Dangara -III (D-III) land owned	Total Dangara land owned (I+II+III)
No. of persons possessed	11	11	25	**47**
Extent of area covered	49.53	44.58	76.49	**170.6**
Sriramput				
Avg. size of land holding	1.35	7.13	28.84	**11.17**
No. of persons possessed	11	17	10	**38**
Extent of area covered	14.88	121.26	288.4	**424.54**
Mania				
Avg. size of land holding	4.18	6.62	8.59	**6.37**
No. of persons possessed	44	38	39	**121**
Extent of area covered	183.79	251.46	335.13	**770.38**
Aligaon				
Avg. size of land holding	2.68	2.74	3.85	**2.92**
No. of persons possessed	47	43	20	**110**
Extent of area covered	126.18	118.06	77.08	**321.32**
Total				
Avg. size of land holding	**2.65**	**4.08**	**5.33**	**3.89**
No. of persons possessed	**346**	**249**	**267**	**862**
Extent of area covered	**918.17**	**1015.71**	**1422.10**	**3355.98**

Source: Field survey 2008-2009

Overall data of Koraput district shows that average size of Dangara-I is 2.65 acres, Dangara-II 4.08 acres and Dangara-III 5.33 acres. The total extent of area of Dangara-I, Dangara-II and Dangara-III is 918.17 acres, 1015.71 acres and 1422.10 acres respectively. In respect of Rayagada district, out of the total Dangara type of land, 92 households are having ownership over total area of 163.18 acre of Dangara-I land with an average size of 1.77 acres. Eighty-nine (89) households occupy 225.51 acres of Dangara-II land having average of 2.53 acres of land and 140 families occupy 292.67 acres of Dangara-III land with average of 2.09 acres of land.

Table 5.2 (b): Dangara type land in Rayagada district (in acres)

Village name	Dangara-I (D-I) land owned	Dangara-II (D-II) land owned	Dangara -III (D-III) land owned	Total Dangara land owned (I+II+III)
Garadi				
Avg. size of land holding	0.0	0.0	1.56	**1.56**
No. of persons possessed	0	0	23	**23**
Extent of area covered	0	0	36	**36**
Chanchadaguda				
Avg. size of land holding	1.0	0.0	1.15	**1.13**
No. of persons possessed	2	0	13	**15**
Extent of area covered	2	0	15	**17**
Bhatapur				
Avg. size of land holding	3.56	4.40	2.48	**3.64**
No. of persons possessed	20	28	17	**65**
Extent of area covered	71.23	123.31	42.27	**236.81**
Majhihalma				
Avg. size of land holding	1.35	2.12	0.95	**1.84**
No. of persons possessed	1	8	2	**11**
Extent of area covered	1.35	17	1.9	**20.25**
Gujalpadu				
Avg. size of land holding	1.25	1.93	2.45	**2.12**
No. of persons possessed	2	19	16	**37**
Extent of area covered	2.5	36.6	39.2	**78.3**
K. Dengasargi				
Avg. size of land holding	2.0	2.0	3.0	**2.69**
No. of persons possessed	2	7	20	**29**
Extent of area covered	4	14	60	**78**
Gunakhal				
Avg. size of land holding	0.0	2.0	2.17	**2.15**
No. of persons possessed	0	3	26	**29**
Extent of area covered	0	6	56.5	**62.5**
Bariguda				
Avg. size of land holding	0.0	3.0	2.0	**2.2**
No. of persons possessed	0	5	20	**25**
Extent of area covered	0	15	40	**55**

Village name	Dangara-I (D-I) land owned	Dangara-II (D-II) land owned	Dangara -III (D-III) land owned	Total Dangara land owned (I+II+III)
Matikona				
Avg. size of land holding	2.45	3.0	0.0	2.48
No. of persons possessed	21	1	0	22
Extent of area covered	51.5	3	0	54.5
Laxmipur				
Avg. size of land holding	0.0	0.0	0.0	0.00
No. of persons possessed	0	0	0	0
Extent of area covered	0	0	0	0
Nathama				
Avg. size of land holding	0.92	1.0	0.0	0.94
No. of persons possessed	3	1	0	4
Extent of area covered	2.75	1	0	3.75
Tadama				
Avg. size of land holding	0.68	0.56	0.6	0.64
No. of persons possessed	41	17	3	61
Extent of area covered	27.85	9.6	1.8	39.25
Total				
Avg. size of land holding	**1.77**	**2.53**	**2.09**	**2.12**
No. of persons possessed	**92**	**89**	**140**	**321**
Extent of area covered	**163.18**	**225.51**	**292.67**	**681.36**

Source: Field survey 2009-2010.

LAND HOLDING PATTERN OF PADDY LAND

Tables 5.3 (a) and 5.3 (b) present the data on landholding pattern of Paddy land in Koraput and Rayagada districts respectively. In case of Koraput district, overall 575 persons possess paddy type lands measuring 826.34 acres with average size of 1.44 acres as shown in Figure 5.3 (a). The land holding pattern of Rayagada district shows that 204 persons are having average size of 3.53 acres covering a total of 720.42 acres of paddy land. [Shown in Fig.5.3 (b)] While comparing between Dangara and Paddy land, the extent of area covered in case of paddy land is less i.e. 482.35 acres of Paddy-I, 221.35 acres of Paddy-II and 122.64 acres of Paddy-III land respectively.

Table 5.3 (a): Paddy type land in Koraput district

Village name	Paddy-I land owned	Paddy-II land owned	Paddy-III land owned	Total Paddy land owned (I+II+III)
Devighat				
Avg. size of land holding	1.90	1.17	1.27	1.67
No. of persons possessed	31	11	4	46
Extent of area covered	58.93	12.9	5.1	76.93
Dhaudapadar				
Avg. size of land holding	1.99	3.13	1.68	2.37
No. of persons possessed	9	10	6	25
Extent of area covered	17.94	31.34	10.09	59.37
Machput				
Avg. size of land holding	3.89	2.39	2.13	3.07
No. of persons possessed	10	11	1	22
Extent of area covered	38.98	26.37	2.13	67.48
Nigamaniguda				
Avg. size of land holding	0.90	3.38	2	2.19
No. of persons possessed	7	8	3	18
Extent of area covered	6.33	27.07	6	39.40
Bhitaragada				
Avg. size of land holding	0.74	0.29	1.59	0.90
No. of persons possessed	26	3	9	38
Extent of area covered	19.27	0.86	14.29	34.42
Baraguda				
Avg. size of land holding	1.80	0.0	0.0	1.80
No. of persons possessed	8	0	0	8
Extent of area covered	14.44	0.0	0.0	14.44
Panchada				
Avg. size of land holding	1.04	1.26	0.99	1.10
No. of persons possessed	54	37	25	116
Extent of area covered	56.44	46.54	24.85	127.83
Jhodipadar				
Avg. size of land holding	0.97	2.1	0.3	1.27
No. of persons possessed	5	3	1	9
Extent of area covered	4.85	6.3	0.3	11.45
Mukhibadai				
Avg. size of land holding	1.35	2.21	1.15	1.50

Village name	Paddy-I land owned	Paddy-II land owned	Paddy-III land owned	Total Paddy land owned (I+II+III)
No. of persons possessed	10	4	4	**18**
Extent of area covered	13.51	8.83	4.62	**26.96**
Sriramput				
Avg. size of land holding	2.10	1.12	0.38	**1.75**
No. of persons possessed	11	4	1	**16**
Extent of area covered	23.15	4.47	0.38	**28.00**
Mania				
Avg. size of land holding	1.97	1.46	1.60	**1.75**
No. of persons possessed	24	12	12	**48**
Extent of area covered	47.25	17.54	19.25	**84.04**
Aligaon				
Avg. size of land holding	1.33	1.23	0.99	**1.25**
No. of persons possessed	136	39	36	**211**
Extent of area covered	181.26	47.92	35.63	**264.81**
Total				
Avg. size of land holding	**1.41**	**1.56**	**1.20**	**1.44**
No. of persons possessed	**331**	**142**	**102**	**575**
Extent of area covered	**482.35**	**221.35**	**122.64**	**826.34**

Source: Field survey 2008-2009.

Table 5.3 (b): Paddy type land in Rayagada district

Village name	Paddy-I land owned	Paddy-II land owned	Paddy-III land owned	Total Paddy land owned (I+II+III)
Garadi				
Avg. size of land holding	1.17	2.5	1.93	1.93
No. of persons possessed	3	4	14	21
Extent of area covered	3.5	10	27	40.5
Chanchadaguda				
Avg. size of land holding	1.6	2.25	2.43	2.12

Village name	Paddy-I land owned	Paddy-II land owned	Paddy-III land owned	Total Paddy land owned (I+II+III)
No. of persons possessed	5	4	7	16
Extent of area covered	8	9	17	34
Bhatapur				
Avg. size of land holding	1.57	1.98	0.81	1.48
No. of persons possessed	7	4	4	15
Extent of area covered	10.99	7.92	3.23	22.14
Majhihalma				
Avg. size of land holding	0.0	0.0	0.0	0.0
No. of persons possessed	0	0	0	0
Extent of area covered	0	0	0	0
Gujalpadu				
Avg. size of land holding	7.97	6.10	5.37	6.01
No. of persons possessed	1	22	6	29
Extent of area covered	7.97	134.14	32.24	174.35
K. Dengasargi				
Avg. size of land holding	1.0	11.62	10.09	10.42
No. of persons possessed	1	10	8	19
Extent of area covered	1	116.24	80.72	197.96
Gunakhal				
Avg. size of land holding	20.75	3.75	3.37	4.63
No. of persons possessed	1	4	10	15
Extent of area covered	20.75	15	33.75	69.5
Bariguda				
Avg. size of land holding	0.0	5.84	2.28	2.93
No. of persons possessed	0	2	9	11
Extent of area covered	0	11.68	20.55	32.23
Matikona				
Avg. size of land holding	1.35	0.5	1.5	1.33
No. of persons possessed	30	1	2	33
Extent of area covered	40.56	0.5	3	44.06
Laxmipur				
Avg. size of land holding	3.39	1.0	0.0	3.15
No. of persons possessed	9	1	0	10
Extent of area covered	30.5	1	0	31.5

Village name	Paddy-I land owned	Paddy-II land owned	Paddy-III land owned	Total Paddy land owned (I+II+III)
Nathama				
Avg. size of land holding	2.31	1.0	1.4	2.00
No. of persons possessed	16	2	5	23
Extent of area covered	37.08	2	7	46.08
Tadama				
Avg. size of land holding	1.5	3.52	0.0	2.34
No. of persons possessed	7	5	0	12
Extent of area covered	10.5	17.6	0	28.1
Total				
Avg. size of land holding	**2.13**	**5.51**	**3.45**	**3.53**
No. of persons possessed	**80**	**59**	**65**	**204**
Extent of area covered	**170.85**	**325.08**	**224.49**	**720.42**

Source: Field survey 2009-2010

The data of Rayagada district in Table 5.3 (b) shows that 80 households were occupying 170.85 acres of paddy-I type of land with 2.13 acres of average size, 59 households having 325.08 acres of paddy-II land with an average size of 5.51 acres and 65 households are having 224.49 acres of paddy-III type land with 3.45 acres of average size.

LAND HOLDING PATTERN OF ATALA LAND

Table 5.4 (a): Atala type land in Koraput district

Village name	Atala-I land owned	Atala-II land owned	Atala-III land owned	Total Atala land owned (I+II+III)
Devighat				
Avg. size of land holding	0.0	0.0	0.0	0.0
No. of persons possessed	0	0	0	0
Extent of area covered	0	0	0	0

Village name	Atala-I land owned	Atala-II land owned	Atala-III land owned	Total Atala land owned (I+II+III)
Dhaudapadar				
Avg. size of land holding	0.0	0.0	0.0	0.0
No. of persons possessed	0	0	0	0
Extent of area covered	0	0	0	0
Machhput				
Avg. size of land holding	0.0	1.78	0.0	1.78
No. of persons possessed	0	3	0	3
Extent of area covered	0	5.35	0	5.35
Nigamaniguda				
Avg. size of land holding	0.2	0.0	0.0	0.2
No. of persons possessed	1	0	0	1
Extent of area covered	0.2	0	0	0.2
Bhitaragada				
Avg. size of land holding	0.0	0.0	0.0	0.0
No. of persons possessed	0	0	0	0
Extent of area covered	0	0	0	0
Baraguda				
Avg. size of land holding	0.0	0.0	0.0	0.0
No. of persons possessed	0	0	0	0
Extent of area covered	0	0	0	0
Panchada				
Avg. size of land holding	2.70	0.0	0.0	2.70
No. of persons possessed	0	0	0	0
Extent of area covered	5.41	0	0	5.41
Jhodipadar				
Avg. size of land holding	0.94	0.0	0.0	0.94
No. of persons possessed	5	0	0	5
Extent of area covered	4.72	0	0	4.72
Mukhibedai				
Avg. size of land holding	0.0	0.0	0.0	0.0
No. of persons possessed	0	0	0	0
Extent of area covered	0	0	0	0
Sriramput				
Avg. size of land holding	0.0	0.0	0.0	0.0

Village name	Atala-I land owned	Atala-II land owned	Atala-III land owned	Total Atala land owned (I+II+III)
No. of persons possessed	0	0	0	0
Extent of area covered	0	0	0	0
Mania				
Avg. size of land holding	2.75	2.52	2.06	**2.56**
No. of persons possessed	6	3	2	**11**
Extent of area covered	16.53	7.56	4.13	**28.22**
Aligaon				
Avg. size of land holding	1.24	1	0.0	**1.21**
No. of persons possessed	18	2	0	**20**
Extent of area covered	22.28	2	0	**24.28**
Total				
Avg. size of land holding	**1.53**	**1.86**	**2.06**	**1.70**
No. of persons possessed	**30**	**8**	**2**	**40**
Extent of area covered	**49.14**	**14.91**	**4.13**	**68.18**

Source: Field survey 2008-2009

Tables 5.4 (a) and 5.4 (b) present the data on landholding pattern of Atala land in Koraput and Rayagada district respectively. Possession of Atala type of land is reflected in Table 5.4 (a) and (b). Majority households of the villages have not possessed this type of land. As a whole, only 40 persons are having Atala type land with average size of 1.70 acres covering a total of 68.18 acres. Villages like Devighat, Dhaudapadar Bhitaragada, Baraguda, Mukhibedei and Srirampur do not possess any kind of Atala lands.

Table 5.4 (b): Atala type land in Rayagada district

Village name	Atala-I land owned	Atala-II land owned	Atala-III land owned	Total Atala land owned (I+II+III)
Garadi				
Avg. size of land holding	1.0	1.0	1.08	**1.04**

Village name	Atala-I land owned	Atala-II land owned	Atala-III land owned	Total Atala land owned (I+II+III)
No. of persons possessed	5	7	13	**25**
Extent of area covered	5	7	14	**26**
Chanchadaguda				
Avg. size of land holding	1.17	1.5	1.0	**1.27**
No. of persons possessed	6	4	1	**11**
Extent of area covered	7	6	1	**14**
Bhatapur				
Avg. size of land holding	2.0	2.5	0.0	**2.44**
No. of persons possessed	1	8	0	**9**
Extent of area covered	2	20	0	**22**
Majhihalma				
Avg. size of land holding	0.0	0.0	0.0	**0.00**
No. of persons possessed	0	0	0	**0**
Extent of area covered	0	0	0	**0**
Gujalpadu				
Avg. size of land holding	0.0	3.95	0.0	**3.95**
No. of persons possessed	0	8	0	**8**
Extent of area covered	0	31.66	0	**31.66**
K. Dengasargi				
Avg. size of land holding	0.0	5.57	3	**4.28**
No. of persons possessed	0	2	2	**4**
Extent of area covered	0	11.14	6	**17.14**
Gunakhal				
Avg. size of land holding	0.0	2.12	1.65	**1.76**
No. of persons possessed	0	4	13	**17**
Extent of area covered	0	8.5	21.5	**30**
Bariguda				
Avg. size of land holding	0.0	1.26	1.51	**1.38**
No. of persons possessed	0	7	6	**13**
Extent of area covered	0	8.86	9.1	**17.96**
Matikona				
Avg. size of land holding	3.86	3.0	0.0	**3.84**
No. of persons possessed	53	1	0	**54**
Extent of area covered	204.52	3	0	**207.52**

Village name	Atala-I land owned	Atala-II land owned	Atala-III land owned	Total Atala land owned (I+II+III)
Laxmipur				
Avg. size of land holding	2.80	1.0	0.0	2.73
No. of persons possessed	23	1	0	24
Extent of area covered	64.5	1	0	65.5
Nathama				
Avg. size of land holding	1.97	1.36	1.0	1.75
No. of persons possessed	29	11	3	43
Extent of area covered	57.12	15	3	75.12
Tadama				
Avg. size cf land holding	8.78	1.58	2.29	7.22
No. of persons possessed	25	6	1	32
Extent of area covered	219.47	9.46	2.29	231.22
Total				
Avg. size of land holding	**3.94**	**2.06**	**1.46**	**3.07**
No. of persons possessed	**142**	**59**	**39**	**240**
Extent of area covered	**559.61**	**121.62**	**56.89**	**738.12**

Source: Field survey 2009-2010

It is found from Table 5.4 (b) that 240 households occupied a total of 738.12 acres of Atala type lands which shows that the average size of Atala land holding is 3.07acres. Of the total Atala lands, Atala-I type of land is possessed by 142 households covering 559.61 acres with 3.94 acres of average size, 59 households with 121.62 acres of Atala-II land having 2.06 acres of average size and 39 households possess 56.89 acres of Atala-III land with 1.46 acres of average size of land.

From the above analysis it can be inferred that in absolute terms, the households possess mostly Dangara type of lands.

LAND ENCROACHMENT CASES

As discussed earlier, households of both the districts possess *patta* land (*raiyat* and lease) and some of the households also possess encroached government land. The Orissa Prevention of Land Encroachment Act, 1972 (OPLE, 1972), governs eviction or regularization of encroachments on government land. As per the Act, unauthorized occupation of government land is illegal. The central purpose of the Orissa Prevention of Land Encroachment Act, 1972 is to prevent unauthorized occupation of government land. It is even-handed and clearly lays down penalties for all instances of encroachment followed by eviction. Through a 1982 amendment, the state has recognized that there are different types of encroachers on different types of lands. While all encroachers are required to be evicted from "objectionable" lands, landless encroachers are legally entitled to settlement up to one standard acre of land per family in respect of "non-objectionable" land (i.e. cultivable wasteland). The provision of settlement of government land to the landless is not a "proactive" provision. Even "eligible" encroachers normally referred to as landless cannot "apply" for regularization, as an act of encroachment is considered illegal in the first place. Moreover, in matters of land settlement even the legal process provides ample scope for manipulation and corruption.

The position of encroachment cases during last three years i.e. 2004-05, 2005-06 and 2006-07 in Koraput district and 2005-06 and 2006-07 and 2007-08 in Rayagada district is shown Table 5.5 (a) and Table 5.5 (b) respectively.

While analyzing the status of possession of encroached lands collected during field survey in the study villages of Koraput district as shown in Table 5.6 (a), a total of 337 persons occupied 483.48 acres of encroached land. It is seen that 283 encroachers occupied 432.24 acres of Dangara land, 52 persons occupied 49.52 acres of Paddy land and 2 persons were having 1.72 acres of Atala land. It shows that encroachment of Dangara lands is more in terms of number of persons and area coverage. But in case of Atala land, only in Jhodipadar village, 2 households possessed such lands. Village wise position of encroached land reveals that number of encroachers is more in Bhitaragada and Panchada villages in comparison to other villages. This may be due to proportionately more number of households living in those villages.

Table 5.5 (a): Position of encroachment cases in Koraput district

Year-wise information	No of cases pending at the beginning of the month	No of cases instituted during the year	Total no. of cases for disposal	Cases disposed off	Balance for disposal	No of cases which encroachment settled	No of cases in which eviction made	Amount of assessment realized	Amount of fine/ penalty realized
2004-05	5119	22486	27605	22772	4833	58	22750	142563.98	2827938.82
2005-06	4868	21390	26958	20806	5452	49	20806	135134.60	2396013.65
2006-07	7095	17789	24884	17602	7282	NA	17602	98093.50	3304884.25

Source: Tahasil office, Koraput district

Table 5. 5(b): Position of encroachment cases in Rayagada district

Year	No. of the cases pending at the beginning of the year	No. of the cases instituted during the year	Total no. of the cases for disposal	Total no. of cases disposed of during the year	Balance cases for disposal	No. of cases in which		Amount of assessment & penalty imposed	Amount collected	Amount pending for realization	No. of cases in which properties forfeitted	No. of cases on which const. stayed & finalised
						Settled	Evicted					
1	2	3	4	5	6	7	8	9	10	11	12	13
2005-06	3083	18844	21927	18223	3724	–	18223	2931819.25	2255049.75	1175789	–	–
2006-07	3704	18122	21826	17835	3991	267	17568	3115777.00	2514100.05	601676.95	–	–
2007-08	3991	16861	20852	18081	2771	69	18012	4116621	2781702	1334919	–	–

Source: Annual administrative report (2005-2008)

Under rule 5-B read with Schedule-V of the Orissa Government Land Settlement Rules provision has been made for settlement of Gram Kantha Paramboke, Abadi, Khasmahal and Nazul lands for homestead purposes by the tehsildars on application of the occupants of such lands within a stipulated period of six months since 11.2.2010 or within such period as may be appointed by government from time to time on payment of prescribed salami and unpaid arrear rent for the last ten years in respect of Nazul and Khasmahal lands before settlement of such lands with the occupants. This process of settlement through revenue officials (tehsildar) would entail extra financial burden on the poor tribal occupants of house sites in the rural areas. Instead, this work should be taken up by regular survey and settlement operations by survey and settlement organization.

Table 5.6(b) presents the position of encroachments in the study villages of Rayagada district. Here also encroachment is confined to three types of land. As indicated in the table, 321 households encroached an area of 601.45 acres of land out of which 223 households occupied 403.39 acres of Dangara land, 21 houses encroached 36.06 acres of Paddy type land whereas 77 households encroached 162 acres of Atala land. The village-wise data shows that 42 households of Bhatpur village encroached maximum area with coverage of 111.40 acres whereas the highest number of households (63 nos.) in Tadama village encroached only 41.25 acres.

While comparing the two districts, maximum encroachments in Koraput district were found in Dangara type land followed by Paddy and Atala land. Only two HHs encroached Atala type land, whereas in Rayagada district, maximum encroachments were in respect of Dangara type followed by Atala and Paddy land.

A total of 31 households of study villages received lease *patta* of 75.29 acres out of which 17 households possessed Dangara land, 13 households got Paddy land and only one household received Atala land. The data considered only agricultural land and excluded homestead land.

Table 5.6 (a): Land holding pattern of encroachments in Koraput district

Name of villages	Encroachment of dangara type land		Encroachment of paddy type land		Encroachment of atala type land		Total Encroachment	
	No. of persons	Extent of area in acres	No. of persons	Extent of area in acres	No. of persons	Extent of area in acres	No. of persons	Extent of area in acres
Devighat	04	6.50	02	2.60	-	-	06	9.10
Dhaudapadar	04	6.24	04	8.88	-	-	08	15.12
Machhput	10	65.14	01	2.13	-	-	11	67.27
Nighamaniguda	02	4.54	00	0.00	-	-	02	4.54
Bhitaragada	78	104.83	32	22.97	-	-	110	127.80
Baragada	05	1.65	00	0.00	-	-	05	1.65
Panchada	104	136.53	05	4.80	-	-	109	141.33
Jhodipadar	20	16.47	03	0.90	2	1.72	25	19.09
Mukhidabei	30	59.33	00	0.00	-	-	30	59.33
Sriramput	08	14.00	02	3.44	-	-	10	17.44
Mania	04	6.50	01	0.80	-	-	05	7.30
Aligan	14	10.51	02	3.00	-	-	16	13.51
Total	283	432.24	52	49.52	2	1.72	337	483.48

Source: Field survey 2008-2009

Table 5.6 (b): Land holding pattern of encroachments in Rayagada district

Name of villages	Encroachment of dangara type land		Encroachment of paddy type land		Encroachment of atala type land		Total Encroachment	
	No. of persons	Extent of area in acres	No. of persons	Extent of area in acres	No. of persons	Extent of area in acres	No. of persons	Extent of area in acres
Garadi	20	32.5	01	7.00	16	17.00	37	56.50
Chancharaguda	01	1.00	-	-	-	-	01	1.00
Bhatapur	32	93.34	03	2.06	07	16.00	42	111.40
Majhihaloma	09	18.00	-	-	-	-	09	18.00
Gujalpadu	27	48.80	04	8.00	01	1.00	32	57.80
K. Dengasarigi	28	76.00	04	6.00	03	10.00	35	92.00
Gunakhal	23	46.50	03	3.50	11	22.50	37	72.50
Bariguda	20	46.00	01	4.00	07	12.50	28	62.50
Matikona	02	2.00	04	5.00	25	72.50	31	79.50
Laxmipur	-	-	-	-	06	9.00	06	9.00
Nathama	-	-	-	-	-	-	-	-
Tadama	61	39.25	01	0.50	01	1.50	63	41.25
Total	223	403.39	21	36.06	77	162.00	321	601.45

Source: Field Survey 2009-2010.

Table 5.7(a): Land holding pattern of lease-hold lands in Koraput district

Name of villages	Lease of dangara type land		Lease of paddy type land		Lease of atala type land		Total lease land	
	No. of persons	Extent of area in acres	No. of persons	Extent of area in acres	No. of persons	Extent of area in acres	No. of persons	Extent of area in acres
Devighat	00	0.00	00	0.00	-	-	00	0.00
Dhaudapadar	00	0.00	06	10.09	-	-	06	10.09
Machhput	04	8.33	04	30.57	01	1.00	09	39.9
Nighamaniguda	00	0.00	01	3.00	-	-	01	3.00
Bhitaragada	00	0.00	00	0.00	-	-	00	0.00
Baragada	00	0.00	00	0.00	-	-	00	0.00
Panchada	04	8.00	02	1.50	-	-	06	9.50
Jhodipadar	00	0.00	00	0.00	-	-	00	0.00
Mukhidabei	00	0.00	00	0.00	-	-	00	0.00
Sriramput	02	3.09	00	0.00	-	-	02	3.09
Mania	01	2.71	00	0.00	-	-	01	2.71
Aligan	06	7.00	00	0.00	-	-	06	7.00
Total	17	29.13	13	45.16	01	1.00	31	75.29

Source: Field Survey 2008-09

Table 5.7(b): Land holding pattern of lease-hold lands in Rayagada district

Name of villages	Lease of dangara type land		Lease of paddy type land		Lease of atala type land		Total lease land	
	No. of persons	Extent of area in acres	No. of persons	Extent of area in acres	No. of persons	Extent of area in acres	No. of persons	Extent of area in acres
Garadi	01	1.00	07	14.00	05	5.00	13	20.00
Chancharaguda	01	1.00	13	25.00	06	7.00	20	33.00
Bhatapur	-	-	-	-	-	-	-	-
Majhihaloma	-	-	-	-	-	-	-	-
Gujalpadu	-	-	01	1.00	-	-	01	1.00
K. Dengasarigi	-	-	-	-	-	-	-	-
Gunakhal	01	2.00	-	-	01	2.00	02	4.00
Bariguda	-	-	-	-	-	-	-	-
Matikona	-	-	-	-	-	-	-	-
Laxmipur	-	-	-	-	-	-	-	-
Nathama	04	3.75	-	-	10	10.00	14	13.75
Tadama	-	-	-	-	02	3.80	02	3.80
Total	07	7.75	21	40.00	24	27.80	52	75.55

Source: Field Survey 2009-10

From the Table 5.7 (b), it is revealed that a total of 52 households of study villages of Rayagada district received lease patta covering 75.55 acres out of which 7 households received 7.75 acres of lease of Dangara land, 21 households possessed paddy land covering 40 acres and 24 households received Atala lands covering 27.8 acres. Table 5.8(a) indicates the average size of landholdings in villages of Koraput district.

Table 5.8 (a): Average size of land holdings (in acres) in Koraput district

Village name	Average Size	Total land in acres
Devighat	2.02	106.92
Dhaudapadar	6.03	162.74
Machhput	7.40	236.84
Nighamaniguda	5.40	189.16
Bhitaragada	1.74	158.69
Baraguda	10.81	183.75
Panchada	4.41	1039.98
Jhodipadar	1.32	38.26
Mukhibedei	4.82	197.56
Sriramput	18.10	452.54
Mania	12.98	882.64
Aligan	2.90	610.41
Total	4.92	4259.49

Source: Field Survey 2008-2009.

From Table 5.8(a) it is thus seen that the average size of land holding and extent of area in the study villages of Koraput district is 4.92 acres and 4259.49 acres of land respectively. The average size of land holding is the highest in Sriramput village (i.e. 18.1 acres) followed by Mania (12.98 acres). In terms of total extent of area, Panchada covered the maximum acres of land i.e. 1039.98 acres followed by Mania with 882.64 acres. On the contrary, the lowest quantity of land i.e. only 38.26 acres is occupied in Jhodipadar. In respect of number of households, Jhodipadar has the lowest number and Panchada has the highest number of households. It implies that total extent of area is found to be proportionately more in villages having large number of households.

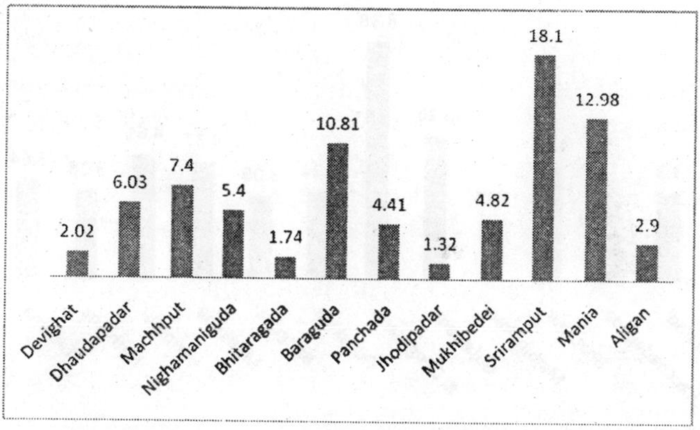

Figure 5.2 (a): Average size of land holdings (in acres) in Koraput district

Table 5.8 (b): Average size of land holdings (in acres) in Rayagada district

Village name	Average size	Total land in acres
Garadi	3.30	102.5
Chanchadaguda	1.80	65
Bhatapur	4.25	280.95
Majhihalma	2.25	20.25
Gujalpadu	5.36	284.31
K. Dengasargi	8.88	293.1
Gunakhal	3.24	162
Bariguda	3.09	105.19
Matikona	4.37	306.08
Laxmipur	4.61	97
Nathama	3.28	124.95
Tadama	3.64	298.57
Total	4.08	2139.9

Source: Field Survey 2009-2010.

Table 5.8 (b) shows that the average size of landholding in the study villages of Rayagada district is 4.08 acres covering 2139.9 acres. K. Dengasargi village is having the highest average size of landholding (8.88 acres) followed by Gujalpadu (5.36 acres) whereas Chancharaguda has the lowest average size of

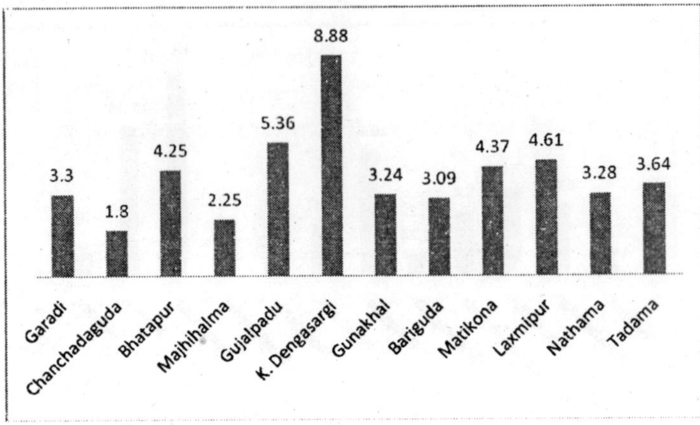

Figure 5.2 (b): Average size of land holdings (in acres) in Rayagada district

landholding (1.80 acres). Maximum area coverage has been revealed in Matikona village (306.08 acres) followed by Tadama village (298.08 acres) as shown in Figure 5.2(b).

OPERATIONAL LAND HOLDINGS

Operational Holdings of Orissa

The total number of holdings for scheduled caste and scheduled tribes in the State in 2000-01 was estimated at 5.69 lakh and 12.30 lakh respectively which showed an increase of 4 per cent and 4.2 per cent over 1995-96. The increase in the number of holdings may be attributed to the division of families as well as cultivation of lands by other castes. Similarly, the average size of holding in the State has decreased from 1.3 hectares in 1995-96 to 1.25 hectares in 2000-01, due to decline in cultivable land and increase in number of holdings. The average size of holding in case of marginal, small, semi-medium, medium and large groups was 0.50 hectares, 1.39 hectares, 2.69 hectares, 5.3 hectares and 16.48 hectares respectively. However, the scheduled castes and scheduled tribes have an average operated area of 0.90 hectares and 1.33 hectares respectively. (Agricultural Census 1995-96 and 2000-01)

Size of Operational Holdings in the Study Villages

The size of operational holdings of HHs of study villages may be examined in three broad categories viz. marginal farmers who have up to 2 acres of land, small farmers within 2 to 5 acres and large farmers who possess above 5 acres of land. The position of both the districts of Koraput and Rayagada is indicated in the Tables 5.9 (a) and (b).

Table 5.9 (a): Size of operational holding in acres in Koraput district

Village name	Land holding category			Total
	Marginal (up to 2 acres)	Small (2.1 to 5.0 acres)	Large (above 5.0 acres)	
Devighat	39 (73.58)	12 (22.64)	2 (3.78)	53 (100.0)
Dhaudapadar	15 (55.56)	1 (3.70)	11 (40.74)	27 (100.0)
Machhput	15 (46.87)	4 (12.5)	13 (40.63)	32 (100.0)
Nighamaniguda	10 (28.57)	13 (37.15)	12 (34.28)	35 (100.0)
Bhitaragada	68 (74.72)	12 (13.19)	11 (12.09)	91 (100.0)
Baraguda	5 (29.41)	0 (0.0)	12 (70.59)	17 (100.0)
Panchada	129 (54.66)	59 (25.0)	48 (20.34)	236 (100.0)
Jhodipadar	24 (82.76)	4 (13.79)	1 (3.45)	29 (100.0)
Mukhibedei	18 (43.90)	14 (34.15)	9 (21.95)	41 (100.0)
Srirampüt	14 (56.0)	2 (8.0)	9 (36.0)	25 (100.0)
Mania	20 (29.41)	8 (11.77)	40 (58.82)	68 (100.0)
Aligan	156 (74.28)	27 (12.86)	27 (12.86)	210 (100.0)
Total	513 (59.36)	156 (18.05)	195 (22.57)	864 (100.0)

Note: Figures in the bracket indicate percentage to the total
Source: Field Survey 2008-2009.

Table 5.9 (a) depicts the data on the size of operational holdings of households of study villages of Koraput district. On the whole, 59 per cent are marginal farmers, 18 per cent small landholders and 22 per cent come under large landholders. Maximum landholders in Jhodipadar (82.76 per cent) are marginal landholders followed by Bhitaragada (74.72 per cent) and Aligaon (74.28 per cent) as indicated in Figure 5.3 (a). The highest percentage of households of Baraguda (70.6 per cent) village possess more than 5 acres of land followed by Mania village (58.82 per cent). On the contrary, very less number of households i.e. only about 4 per cent of Jhodipadar village are large landholders. The highest number of small landholders possessing between 2.1 to 5 acres of land is found in Nighamaniguda (37 per cent) followed by 34 per cent in Mukhibedei village.

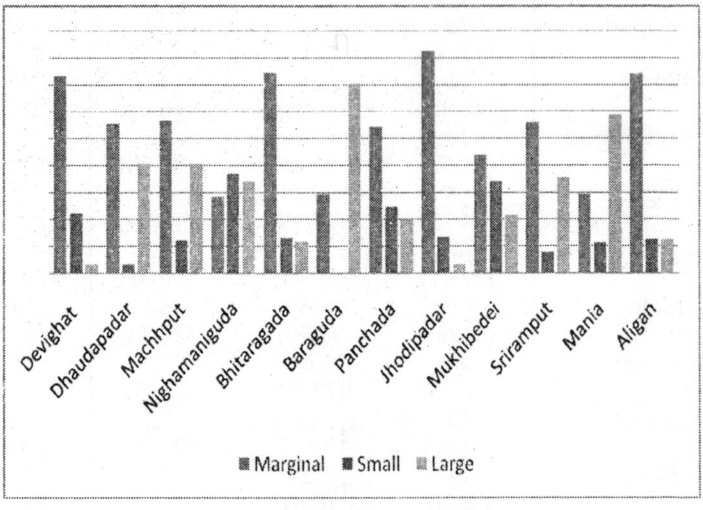

Figure 5.3 (a): Size of operational holding in acres in Koraput district

Study reveals that in regard to the size of the operational holding in the study villages of Rayagada district [Table 5.9 (b)], out of 916 total households, 523 households possess land. Among the landholders, 54.68 per cent are small farmers, 26.39 per cent are marginal farmers and 18.93 per cent are large farmers indicated in Fig. 5.3(b). In Tadama village 63 (76.83

per cent) households are marginal farmers whereas Garadi has the lowest 8 (25.80) per cent of marginal farmers. Likewise, Garadi village is having the highest 20 (64.52 per cent) percentage of small farmers and Tadama the lowest (10.77 per cent).

Table 5.9 (b): Size of operational holding in acres in Rayagada district

Village name	Land holding category			Total
	Marginal (up to 2 acres)	Small (2.1 to 5.0 acres)	Large (above 5.0 acres)	
Garadi	8 (25.80)	20 (64.52)	3 (9.68)	31 (100.0)
Chanchadaguda	30 (83.33)	5 (13.89)	1 (2.78)	36 (100.0)
Bhatapur	29 (43.94)	25 (37.88)	12 (18.18)	66 (100.0)
Majhihalma	4 (44.44)	5 (55.56)	0 (0.0)	9 (100.0)
Gujalpadu	22 (41.51)	9 (16.98)	22 (41.51)	53 (100.0)
K. Dengasargi	10 (30.30)	8 (24.24)	15 (45.46)	33 (100.0)
Gunakhal	38 (76.0)	7 (14.0)	5 (10.0)	50 (100.0)
Bariguda	26 (76.48)	4 (11.76)	4 (11.76)	34 (100.0)
Matikona	20 (28.58)	33 (47.14)	17 (24.28)	70 (100.0)
Laxmipur	12 (57.14)	4 (19.05)	5 (23.81)	21 (100.0)
Nathama	24 (63.16)	9 (23.68)	5 (13.16)	38 (100.0)
Tadama	63 (76.83)	9 (10.97)	10 (12.20)	82 (100.0)
Total	286 (54.68)	138 (26.39)	99 (18.93)	523 (100.0)

Note: Figures in the bracket indicate percentage to the total
Source: Field survey 2009-2010.

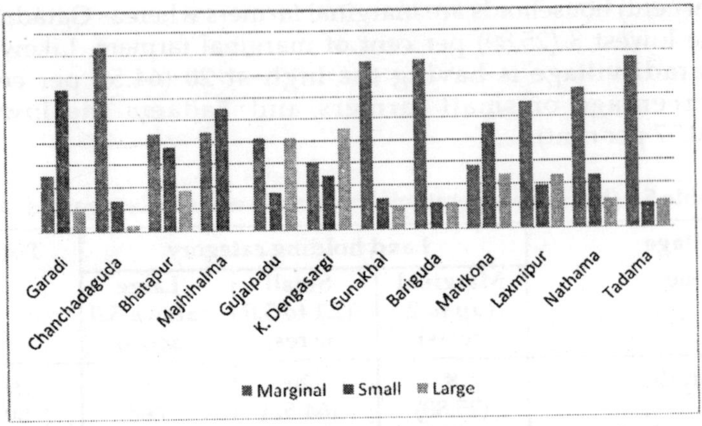

Figure 5.3 (b): Size of operational holding in acres in Rayagada district

It is apparent that, the important aspect of the land holding is its size and quality, which is nucleus of the agricultural production. When the farm is too small, the cultivator is at loss as the relative cost of inputs will be high, improvements will be out of reach and he will not get full return for the labour and money invested. The figures shown above indicate that marginal holdings are more in Koraput district whereas small holdings are found more in Rayagada district. The study shows that the land holdings are by and large small which may be an important factor to indicate the poor condition of the households.

KEY FINDINGS

1. The study has revealed that in relation to the nature of ownership of households of Koraput district, out of 2439 households, 651 HHs constituting 26 per cent HHs of study villages possess land of their own. Six hundred and fifty five (655) HHs constituting 26.16 per cent by inheritance and 1133 HHs constituting 46.45 per cent are landless. This indicates that more than 50 per cent are having land in either way. In case of Rayagada district, out of 916 households, 223 (24.35%) households are having lands in their own possession, 428 households constituting 46.72% possess land by

inheritance and 265 (28.93%) have no possession over land at all. On a comparison between the two districts, it is seen that land holdings are more in Rayagada district than Koraput district.

2. The overall data of Koraput district shows that average size of Dangara-I land is 2.65, Dangara-II 4.08 and Dangara-III is 5.33 acres respectively. A total of 862 persons possessed Dangara land covering a total of 3355.98 acres. The average size of holding of Dangara land is 3.89 acres in Koraput district. The average size of holding is found maximum in Srirampput village (11.17) acres followed by Baraguda (7.05) acres. The maximum area of Dangara land is available in Panchada village (906.74) acres followed by Mania village (770.38). The highest number of persons (292) in Panchada followed by Mania (121) possessed Dangara lands. This might be due to large number of households living in these villages.

   Similarly, in case of Rayagada district, 92 households are having ownership over total area of 163.18 acres of Dangara-I lands with an average size of 1.77 acres and 89 households occupy 225.51 acres of Dangara-II land with 2.53 acres of average size and 140 families occupy 292.67 acres of Dangara-III land having 2.09 acres of average land size. Out of the total 916 households, 321 households are having 2.12 acres of average size of Dangara lands with a total area of 681.36 acres. The highest average size of land holding is found in Bhatpur village (3.64 acres) covering maximum area of 236.81 acres possessed by 65 persons. Average size of holding of Dangara land is more in case of Koraput district than Rayagada district.

3. While comparing between Dangara and Paddy lands in Koraput district the extent of area covered under Paddy land is very less i.e. 482.35 acres of Paddy-I, 221.35 acres of Paddy-II and 122.64 acres of Paddy-III land. Overall data shows that 575 persons possess Paddy type land covering 826.34 acres with the average size of 1.44 acres. In Rayagada district, 80 households occupied 170.85 acres of Paddy-I type of land with 2.13

acres as average size, 65 households with 224.49 acres of Paddy-III type land and 59 households possessing 325.08 acres of Paddy-II land. Considering the overall data, 204 households are having average size of 3.53 acres covering a total of 720.42 acres of Paddy land.

4. As a whole, in Koraput district, 40 households possess Atala type land with an average size of 1.70 acres covering total of 68.18 acres. Similarly, in Rayagada district, 240 households possess 738.12 acres of Atala type of land with average size of 3.07 acres. Average size of holding of Paddy and Atala land is more in Rayagada district than Koraput district.
5. In Koraput district, out of 2439 total sample households, 337 households occupied 483.48 acres of encroached land. Among the encroachers, 283 households occupied 432.24 acres of Dangara land, 52 households 49.52 acres of Paddy land and 2 households were having 1.72 acres of Atala land.
   In case of Rayagada district, 321 persons were encroaching an area of 601.45 acres of land out of which 223 persons were occupying 403.39 acres of Dangara land, 21 persons encroached 36.06 acres of Paddy type land and 77 persons encroached 162 acres of Atala type land.
6. While analyzing the position of lease land in Koraput district, it is seen that out of total 31 lease land holdings covering 75.29 acres of land, 17 persons received Dangara type land covering 29.13 acres and 13 persons got Paddy type lands covering 45.16 acres. Only one person has one acre of Atala land. Similarly, in Rayagada district, out of 52 total lease patta holders covering 75.55 acres, 7 households received 7.75 acres of Dangara land, 21 households 40 acres of Paddy land and 24 households received 27.80 acres of Atala land.
7. The average size of land holding is 4.92 acres covering total area 4259.49 acres in Koraput district. The highest average size of land holding is found in Srirampur (i.e. 18.1 acres) village followed by Mania (12.98 acres) village. With regard to total extent of area, Panchada village has the highest acres of land i.e. about 1039.98 acres followed by Mania of about 882.64 acres of land

In case of Rayagada district, the average size of land holding is 4.08 acres covering total area of 2139.9 acres of land. The highest average size of land holding is found in K. Dengasargi (8.88 acres) village whereas Chancharaguda village has the lowest average size of land (1.80 acres). Considering the total size of land holding, Matikona village has the maximum area (306.08 acres) and Chancharguda the minimum (65 acres) area.

8. Regarding the operational size of land holding of Koraput district, 59 per cent are marginal farmers, 18 per cent small landholders and 22 per cent are large landholders. Maximum number of landholders are in Jhodipadar (82.76%) as marginal landholders followed by Bhitaragada (74.72%) and Aligaon (74.28%) villages. Highest percentage of households is found in Baraguda (70.6%) village which possess more than 5 acres of land followed by Mania (58.82%) village. In Rayagada district, 54.68 per cent of the households are small farmers, 26.39 per cent of the households marginal farmers and 18.93 per cent are large farmers. In Tadama village, the highest percentage (76.83%) of householders are marginal farmers whereas Garadi village has the lowest i.e. 25.8 per cent number of marginal farmers. Likewise, the highest (64.52%) in Garadi are found to be small farmers and the lowest (10.77%) in Tadama village. The highest percentage of households (45.46%) is found in K.Dengasargi village. Between the two districts, marginal farmers are found more in Koraput district than Rayagada district.

6

Land Alienation and Restoration

This chapter focuses on types of land alienation, method of transfer and causes of transfer of tribal land in the study villages. This chapter also deals with restoration of tribal land under Orissa Scheduled Areas Transfer of Immovable Properties (OSATIP), 1956 Regulation (2) followed by 2002 Amendment.

Alienation of tribal land has acquired tremendous dimension in the past and continues to acquire equally critical assumptions even in the present day conditions. As tribals continued to live in comparative isolation during the British Raj, the problems of land in the tribal tracts remained neglected because of inaccessibility of tribal regions to the main streams of administration both during the 18th and first half of 19th century. It is only during the second half of 19th century and the early 20th century that these areas were gradually developed due to developmental interventions. Due to growth of population and expansion of economic activities and inter-regional movement of non-tribal population, the economic and administrative structures enhanced the degree of security to existing trade, commerce and industrial activities. In consequence, the law and order mechanism offered a measure of security to the infiltrators like petty traders and money-lenders.

Historically, alienation of land from tribals has been recognized as an important political and administrative issue. Protection of land rights of tribal communities was incorporated into Indian constitution through Schedule-V and VI. Legally, alienation of land is defined as loss of *patta* of private land. Loss of access to land and forests classified as state owned and diverted for various purposes, have rarely figured in the official discourse of tribal land alienation. But access to land is of fundamental importance in rural/tribal India. Previous studies show that trbals lost their land due to illiteracy, ignorance and awareness about their rights and entitlements. In this connection, the present study tried to assess the degree of awareness about different land related Acts/laws especially laws preventing transfer of tribal lands as indicated in Table 6.1(a) and 6.1(b).

UNDERSTANDING OF GENERAL PUBLIC IN THE STUDY AREAS ABOUT ACTS/LAWS

The awareness of people about different laws, rules and regulations and their capacity with the willingness to use the available facilities is one of the indicators of social development that would confirm the impact of developmental efforts made by the government. The present study sought to enquire whether the people in study villages are aware of the existing Acts/laws. It is learnt from research studies that tribals are normally ignorant of land related Acts/laws as a result of which they are deprived of their entitlements. Table 6.1(a) represents the data on awareness about the land related Acts/laws in Koraput district. On the whole, about 53 per cent of people have some idea about few land related Acts mostly land passbook and "Mo Jami Mo Diha". Very few have knowledge about PESA Act even though they belong to Scheduled Areas. Study results reveal that tribals lack awareness because they were not provided with any training in regard to the Acts/laws prior to ISS intervention in the study villages.

Table 6.1(a): Awareness about the land related Acts/laws in Koraput district

Name of the village	No. of respondents	Awareness/ understanding	Awareness about different laws						
			1	2	3	4	5	6	7
Devighat	204	74(36.3)	29	1	16	14	17	73	74
Dhaudapadar	132	11(8.3)	9	0	5	2	2	11	11

Name of the village	No. of respon-dents	Awareness/ underst-anding	Awareness about different laws						
			1	2	3	4	5	6	7
Machhput	66	14(21.2)	6	0	7	3	4	13	11
Nighamaniguda	145	49(33.8)	28	3	22	12	14	47	46
Bhitaragada	143	64(44.7)	1	1	15	38	41	62	64
Baraguda	46	18(39.1)	0	0	9	16	18	18	18
Panchada	569	321(56.4)	132	5	95	57	113	321	318
Jhodipadar	42	19(45.2)	18	1	17	4	9	17	18
Mukhibedei	232	197(84.9)	136	13	120	35	90	197	197
Srirampu t	85	79(92.9)	35	6	44	37	35	77	79
Mania	331	254(76.7)	76	7	58	19	88	254	254
Aligan	444	201(45.3)	95	53	49	58	36	183	198
Total	2439	1301(53.3)	565	90	457	295	467	1273	1288

Note: Figures in the bracket indicate percentage to the total
Source: Field survey 2008-2009

Note: 1.Awareness about OLR; 2. Awareness about PESA; 3. Awareness about Regulation (II) of 1956; 4. Awareness about Orissa Land Acquisition Act (1986); 5. Awareness about Orissa Survey and Settlement Act (1958); 6. Awareness about Orissa Land Passbook Law; 7. Awareness about "Mo Jami Mo Diha".

Interacting with people in a group while conducting FGD, it has been found thai except Devighat, Machhput and Baraguda villages in the study area of Koraput district, people of no other village seemed to have any knowledge about the Acts/rules/ regulations etc. During interaction, they rather suggested to organise more training and awareness programmes on land related Acts/laws in their respective villages.

Table 6.1(b): Awareness about the land related Acts/laws in Rayagada district

Name of the village	No. of respon-dents	Awareness/ underst-anding	Awareness about different laws						
			1	2	3	4	5	6	7
Garadi	42	19 (45.2)	2	0	19	0	0	0	1

Name of the village	No. of respondents	Awareness/ understanding	Awareness about different laws						
			1	2	3	4	5	6	7
Chanchadaguda	71	39 (54.9)	5	0	38	0	0	1	3
Bhatapur	125	42 (33.6)	42	18	0	0	0	0	0
Majhihalma	36	24(66.7)	24	1	0	0	0	0	0
Gujalpadu	74	3(4.05)	3	3	1	1	1	0	0
K. Dengasargi	43	0(0.0)	0	0	0	0	0	0	0
Gunakhal	80	0(0.0)	0	0	0	0	0	0	0
Bariguda	47	0(0.0)	0	0	0	0	0	0	0
Matikona	109	15(13.8)	0	0	15	0	0	0	0
Laxmipur	55	7(12.7)	0	0	7	0	0	0	0
Nathama	91	1(1.1)	1	1	1	0	0	0	0
Tadama	143	135(94.4)	135	12	4	1	0	1	2
Total	916	285(31.1)	212	35	85	2	1	2	6

Note: Figures in the bracket indicate percentage to the total

Source: Field survey 2009-2010

Note: 1. Awareness about OLR; 2. Awareness about PESA; 3. Awareness about Regulation (II) of 1956; 4. Awareness about Orissa Land Acquisition Act (1986); 5. Awareness about Orissa Survey and Settlement Act (1958); 6. Awareness about Orissa Land Passbook Law; 7. Awareness about "Mo Jami Mo Diha".

Table 6.1 (b) shows the data about knowledge of the people of Rayagada district on different land related Acts/laws. Only 31 per cent people of study village of Rayagada district were aware of land related Acts/Rules, Regulations etc. Among the study villages, in Tadama village, as per household data, 95 per cent were having some knowledge about land related Acts whereas only one person in Nathama village has some idea about different land related Acts/Rules, Regulations etc. While interacting with people during FGD, it was found that the villagers of Bhatpur village are aware about the rules and regulations with regard to the lands. People of Gunakhal and Chanchuraguda villages have no knowledge about land related

Acts and provisions except knowing that they cannot transfer or sale their lands to any non-tribal.

LAND ALIENATION—CONCEPT AND MEANING

The various conformity legislations in different tribal states in India have proved that the spirit of social welfare legislation can be totally flawed by carefully selecting words and phrases in law that kills the soul while maintaining the body of legislation. The law on tribal self-rule which recognized for the first time the competence of a village assembly to manage its community resources, and which also recognized for the first time that a village where one resides is not always a homogeneous, population-based entity but a social cohesive unit with its own self-identity where people who have been ordinarily and traditionally residing for centuries with a common belief system and cultural traits apart from the manner in which they manage their natural resources. Despite these objectives, the states having scheduled areas have proved that it is too difficult to relinquish power in a bureaucratic power structure. But perhaps, it is too late for states to undermine the significance of communities living close to natural resources on which they depend.

Technical meaning of land transfer refers to three concepts i.e. *transaction, acquisition* and *alienation*. Transaction is the sale, purchase and mortgage of land. Acquisition of land is the transfer of private land for the purpose of Government projects/companies etc. If government land is transferred for Government purpose that is called land alienation. *The present study is restricted to the meaning of land alienation*. It is the transfer of ownership of land. Broadly they are (i) land acquisition, and (ii) land alienation including sale, *benami* transfers, mortgage, forcible occupation by non-tribals, oral transfers, transfer to wives, transfer for public purposes and even lands transferred by way of encroachments.

LAND ACQUISITION

Land Acquisition is a method of land alienation where land is transferred from individual/group to government for development projects etc. Orissa is extremely rich in minerals most of which are based in tribal districts. The hilly terrain and availability of water makes it suitable for construction of reservoirs and dams. The National Planning Commission estimates that some 50 million people have been displaced since 1950 on account of various development projects, of which more that 40 per cent are tribals. These projects include large irrigation dams, hydroelectric projects, open-cast and underground coal mines, super thermal power plants and mineral-based industrial units. In the name of development, tribals are displaced from their traditional habitats and livelihoods with little or no rehabilitation. These tribals are pushed into unemployment, debt bondage and hunger due to loss of access to traditional sources of livelihood viz., land, forests, rivers, pastures, cattle etc. This has led to conflicts between the local inhabitants, especially tribals, and the companies that push for development projects in tribal areas and the government machinery. Such incidents have taken place in recent years in Kalinga Nagar, Kashipur, Lanjigarh, and Paradip in Orissa. Although land acquisition is the main reason for conflict, there are other reasons like inadequate rehabilitation and resettlement measures, and perceived state protection of the interests of the companies and industries. Although land transfer and the subsequent alienation of tribal land has been restricted by legislation as a matter of public policy, currently the Government has only achieved limited success in addressing the problems for which they were promulgated. Land alienation, especially in the scheduled areas, both institutional and non-institutional, involves a wide range of issues starting from land use change due to mining to change of ownership due to indebtedness. Institutional factors that cause displacement through transfer of *patta* land are visible. The non-institutional factors through transfers and mortgages at the individual levels are mostly invisible, and are short-term generally removed from public knowledge/domain. This makes it difficult to assess and estimate the extent of its prevalence (Status Report, UNDP, 2008). As per an estimate, less than 25 per cent of the displaced tribals were ever resettled even partially. The casualness with which displacement of tribals has been treated is evident by the fact that out of 13 major dam projects before 1990, no data relating to displaced scheduled tribes is available for seven projects (Study,

Ekta Parishad). As shown in Table 6.2, total area acquired for conservation projects is 1215679 hectares affecting 2170 villages. Table 6.2 indicates the area acquired and number of villages affected by development and conservation projects in Orissa.

Table 6.2: Area acquired and number of villages affected by development and conservation projects in Orissa

Projects	Area acquired or affected (ha.)	No. of villages affected
Irrigation- Dam Projects (Major) Completed.	200423	900
Irrigation - Dam Projects (Medium) Completed	14403	118
Irrigation - Dam Projects (Medium) Ongoing	12160	92
Irrigation - Dam Projects (Major) Proposed	30233	113
Industries (All Types of Industries)	45358	176
Mines (All types of mine leases given out)	101947	NA
Wildlife Sanctuaries and National Parks	811155	771
Total	1215679	2170

Source: "Our Enduring Struggle" a published Booklet by EKTA Parishad.

Complete data about land-acquired villages, villages displaced or affected by various development projects has not been compiled properly till date. The data as provided in the table may be taken as the lowest estimate and it does not include people affected by minor irrigation projects, infrastructure projects such as roads etc. List of a few major irrigation projects in Koraput district areas of Orissa is given in Table 6.3.

Table 6.3: Irrigation Projects in Koraput district

Name	Area (in hac.)	No. of HHs	No. percentage of tribal HHs affected
Machhkund	6477	2938	1500 (51.0)
Balimela	17496	1900	1507 (79.3)
Upper Kolab	11350	3179	1421 (44.7)
Indrawati	17263	5534	2324 (42.0)

Source: "Our Enduring Struggle" a published booklet by EKTA Parishad.

Except for a few irrigation projects, developmental projects have not provided land as compensation. Even where the principle of land for land compensation was accepted, often cash compensation was given, as suitable land was not available. A study of seven development projects with a sample of 301 HHs (with 43.8 per cent tribal households within the sample) showed that legal landlessness increased from 15.6 per cent of the households to 58.8 per cent, after displacement (Pandey, 1998). More important is that since large areas of land cultivated by scheduled tribes were not legally settled in their names, they received no compensation when such land was taken up for development projects. A.B. Ota, in his study of displacement in Upper Indravati Project found that on an average, each displaced family had been cultivating 1.50 acres of state owned and 2.34 acres of private land before displacement and that 49 per cent of the sampled families were landless. After displacement, landlessness increased to 85.25 per cent, the average legal landholding declined to 0.62 acres and the average government land cultivated came down to only 0.2 acres. (Ota, 2001). Table 6.4 indicates the data of some Irrigation and Industrial Projects in scheduled areas of Orissa.

Table 6.4: Some irrigation and industrial projects in scheduled areas of Orissa

Name of the project area	Area (in ha)	No. of HHs affected	Tribal HHs as % of total HHs affected
Machhkund	6477	2938	51
Salandi	1229	589	93
Balimela	17496	1900	79
Upper Kolab	11350	3179	44.7
Indrawati	17263	5534	42
Rourkela Steel Plant	7917	2367	70
HAL, Sunabeda	3764	468	NA
NALCO, Damanjodi	3444	788	50.5
M/s. Orissa Sponge Iron Ltd.,-Palspanga	176	NA	NA
IPITATA, Joda	129	NA	NA
UAIL, Kashipur	862	NA	NA
Vedanta Alumina Ltd., Lanjigarh	707	NA	NA
M/s. Jindal Steel & Power Ltd.	94	NA	NA
Total	70908		

Source: "Our Enduring Struggle" a published Booklet by EKTA Parishad

As regards land acquisition, in study villages of Koraput district, it is seen that lands were acquired mainly for Upper Kolab Project, Hindustan Aeronautics Ltd., Sunabeda and NALCO, Damanjodi. Compensation for acquiring land was made under the legal provisions of Land Acquisition Act which provides that compensation will be paid according to the prevailing market price. It was land for land or land and/or cash. But in Rayagada district, acquisition cases were not found in study villages. The data on acquisition of land in Koraput district and Rayagada district showing number of cases at the beginning of the year and area involved, number of cases instituted and area involved, number of cases disposed off, amount of compensation paid, average cost per acre and balance area are shown in Table 6.5 (a) and Table 6.5 (b).

In Koraput district, 21 acquisition cases covering 569.32 acres were found in the beginning of the year i.e. 2006-07, but during that year only one case was instituted covering 17.35 acres. Out of total 22 cases 8 cases were disposed off during this year paying a total compensation of Rs. 25205405.00 at the rate of Rs. 343600.00 per acre of acquisition.

In Rayagada district, during the same year, a total of only 3 cases were disposed off with payment of total compensation of Rs. 2267426 at the rate of Rs. 115215.00 per acre of acquisition.

MUTATION CASES

While interacting with the key stakeholders, it was revealed that transfer of land whether through sale or verbal transfer or transfer due to partition, mutation cases are not instituted accordingly. In regard to transfer through sale of land and transfer from one generation to another i.e. from father to son/sons, there is no problem for mutations. But if partition is among more than one generation, it becomes difficult for instituting and disposing off mutation cases. As reported by Tahasildar, Koraput, survey and settlements should be done immediately for updating records. Now it is very difficult on their part to update the mutation cases because the numbers of legal heirs have been increased even more than 50 in a family in some cases. As a result mutation cases cannot be settled even at higher level.

Table 6.5 (a): Acquisition of land in Koraput district

Year-wise information	No. of cases at the beginning of the year & area involved		No. of cases instituted during the year & area involved		Total no. of cases instituted and area involved		No. of cases disposed off & area occupied		Amount of compensation paid (Rs.)	Average cost per acre (Rs.)	Balance	
	No.	Area	No.	Area	No.	Area	No.	Area			No.	Area
1	2	3	4	5	6	7	8	9	10	11	12	13
1996-97	NA	NA	NA	NA	NA	NA	NA	NA	NA	-	NA	NA
1997-98	NA	NA	NA	NA	NA	NA	NA	NA	NA	-	NA	NA
1998-99	–	–	4	17.87	–	–	–	–	2,27,283	12,719	–	–
1999-2000	–	–	-	-	–	–	–	–	-	-	–	–
2000-01	–	–	7	34.25	–	–	–	–	6,84,863	19,996	–	–
2001-02	–	–	1	1.08	–	–	–	–	67,592	67,592	–	–
2002-03	–	–	5	196.95	–	–	–	–	72,85,613	36,992	–	–
2003-04	–	–	-	-	–	–	–	–	-	-	–	–
2004-05	–	–	6	440.51	–	–	–	–	11,62,73,638	2,90,684	–	–
2005-06	–	–	16	136.00	–	–	–	–	10,80,698	-	–	–
2006-07	21	569.32	1	17.35	22	586.67	8	443.58	2,52,05,405	3,43,600	-	-

Source: Annual land revenue administration report, 1996-2007.

Table 6.5 (b): Acquisition of lands in Rayagada district

Year-wise information	No. of cases at the beginning of the year & area involved		No. of cases instituted during the year & area involved		Total no. of cases instituted and area involved		No. of cases disposed off & area occupied		Amount of compensation paid (Rs.)	Average cost per acre (Rs.)	Balance	
	No.	Area	No.	Area	No.	Area	No.	Area			No.	Area
1	2	3	4	5	6	7	8	9	10	11	12	13
1996-97	Nil	Nil	Nil	Nil	Nil	Nil	Nil	Nil	Nil	Nil	Nil	Nil
1997-98	2	21.37	2	21.37	2	21.37	2	21.37	41,573	1945	–	–
1998-99	2	10.61	2	10.61	2	10.61	2	10.61	4,47,698	42,200	–	–
1999-2000	9	75.19	9	75.19	9	75.19	9	75.19	38,70,692	51,478	–	–
2000-01	5	12.66	5	12.66	5	12.66	5	12.66	16,49,690	1,30,307	–	–
2001-02	Nil	Nil	Nil	Nil	Nil	Nil	Nil	Nil	Nil	Nil	Nil	Nil
2002-03	15	65.03	15	65.03	15	65.03	15	65.03	26,94,000	41,427	–	–
2003-04	11	41.49	11	41.49	11	41.49	11	41.49	18,99,807	45,790	–	–
2004-05	1	9.7	1	9.7	1	9.7	1	9.7	2,00,000	20,618	–	–
2005-06	11	95.13	11	95.13	11	95.13	11	95.13	10,652,607	11,979	–	–
2006-07	3	19.61	3	19.61	3	19.61	3	19.61	22,67,426	1,15,215	–	–
2007-08	Nil	Nil	Nil	Nil	Nil	Nil	Nil	Nil	Nil	Nil	Nil	Nil
TOTAL	59		59		59		59		2,37,23,493	4,60,959		

Source: Collectorate, Rayagada-2009-2010.

During the conduct of FGDs, it has been also noticed that the RoRs are not being updated as a lot of money has to be spent for getting RoR or *Patta* and in case of joint *Patta*, the RoRs have not been transferred to their legal heirs because if one of the owners is willing, another does not. So it is held-up. Table 6.6 (a) and (b) indicate the position of mutation cases of last three years of the study i.e. 2004-05,2005-06 & 2006-07 and 2005-06, 2006-07 & 2007-08 in Koraput and Rayagada districts respectively.

Table 6.6 (a): Status of mutation cases in Koraput district

Year-wise information	No of cases as on 1st April	No. of cases instituted from April to March	Total cases for disposal	No. of cases disposed-off from 1st April- - 31st March	Balance of cases for disposal
2004-05	587	9213	9800	9084	716
2005-06	716	6437	7153	6708	445
2006-07	445	7373	7818	6967	851

Source: Annual land revenue administration report, 1996-2007.

Table 6.6 (b): Status of mutation cases in Rayagada district

Year-wise information	No. of cases as on 1st April	No. of cases instituted from April to March	Total cases for disposal	No. of cases disposed-off from 1st April- - 31st March	Balance of cases for disposal
2005-06	2913	6143	9056	5258	3798
2006-07	3798	5176	8974	5691	3283
2007-08	2933	6024	8957	5775	3182

Source: Annual Revenue Administration Report, 2005-08.

LAND ALIENATION

The first and foremost form is the manipulation of land records. The unsatisfactory state of land records contributed a lot to the problem of land alienation. The tribals were never legally recognized as owners of the lands which they cultivated. The second form of land alienation is reported to have taken place due to 'benami' transfers. The report of the study team of the Union Home Ministry (May 1975) pointed out that large scale transfers of ownership of the *adivasi* lands are being allowed to go out of hands through illegal and *benami* transactions, collusive civil proceedings etc., in which land remains to be in the names of the original owners who are reduced to the level of mere share-croppers. Another form of land alienation is related to the leasing or mortgaging of the land. To raise loans for various needs, the tribals have to give their land as mortgage to the local money-lenders or to the rich farmers. Encroachment is another form of dispossessing the tribals of their lands and allowing the new entrants in all the places where there are no proper land records to establish the right over land. Bribing the local Patwari (RI) for manipulating the date of settlement of land disputes and ante-dating records, facts and events are resorted to for establishing claim over tribal lands. Concubine or marital alliance is another form to circumvent the law and grab tribal lands at no cost at all. Pretended adoption of the non-tribals by the tribal families is also another method to snatch away the lands of the tribals. Slackness in the implementation of the restrictive provisions also encourages the non-tribals to occupy the tribal lands.

Indebtedness is also one of the main reasons of tribal land alienation. The tribal people depend largely on private money-lenders for their loans who charge high rates of interest. Unable to pay back the loan amount, they are forced to pledge their lands which can never be repaid because of their poverty. The tribals often default in repayment of either interest or capital and are forced to mortgage their lands which are seized by the money-lenders subsequently. This method of transfer of most of the tribal land is done by mortgage with the intention that when the transfer or accumulated the required amount to redeem the mortgage, the land would be released. But in most of the cases, it has been found from the study that due to exorbitant rate of interest, the tribals were unable to arrange the required amount and therefore, finally had to surrender the land to the money-lenders or to the other buyers to pay off the old debt. The rate of interest being very high and sometimes, the entire repayment amount being adjusted towards the interest keeping the principal intact, the tribals are compelled to mortgage such land and are unable to restore.

Land alienation that takes place in various ways has assumed alarming proportion threatening the right to life of the tribal population. Though the problem lies elsewhere, it is being unfortunately always interpreted as the handiwork of certain individuals like the money-lenders, traders, landlords, etc, without understanding the class connection of these individuals. The unsystematic land records of the pre-colonial and colonial periods were followed and adopted by the present state. There was collection of taxes, a strange phenomenon for the natives and it was almost the beginning process of alienation in the tribal areas.

Land Alienation Cases in Study Area

As reflected in Table 6.7, it is seen that out of 2439 total households in the study villages of Koraput district, a total of 215 cases of transfer involving all categories (8.81%) and 121 cases relating to ST cases (4.96%) had taken place whereas, in case of Rayagada district, out of 916 households, 78 transfer cases (8.51%) in respect of scheduled tribes were effected. The total number of transfer cases was 100 constituting 10.92 per cent. Transfer of ST lands includes transfer between ST and ST, ST and SC and ST and others. Taking the overall data, it was found that out of 3355 households, 9.39 per cent cases related to transfer between all categories of households and 5.93 per cent to transfer of ST lands only. The data shows that transfers have taken place more in Rayagada than Koraput district as shown in Figure 6.1.

Table 6.7: Number of land alienation cases

Name of the district	Total number of households	Total transfers of all categories	Transfers of ST households
Koraput	2439 (100.0)	215 (8.81)	121 (4.96)
Rayagada	916 (100.0)	100 (10.92)	78 (8.51)
Total	3355 (100.0)	315 (9.39)	199 (5.93)

Note: Figures in the bracket indicate percentage to the total.
Source: Field survey 2008-2010.

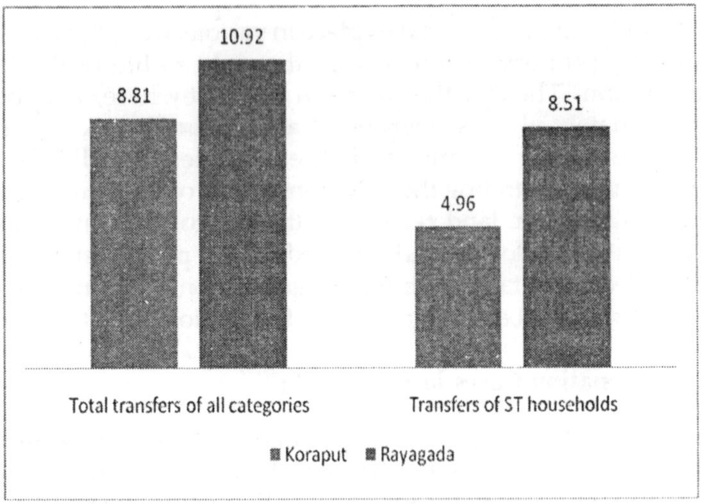

Figure 6.1: Number of land alienation cases

Methods of Land Alienation

As observed during the study, various methods of transfer of land were adopted by different parties. The methods were mainly through sale, mortgage, mortgage with money-lenders, oral transfers, encroachment, public purpose, transfer, land acquisition and cheating. But most of the transfers in tribal areas had taken place through oral commitments. In the absence of settlement records, the tribals had no documentary proof of title over their lands. In certain areas, ownership of land and incidence of tenancy was guided by the 'common will' of the village community. In the circumstances, alienation of tribal land towards repayment of various types of loans was oral which had almost assumed an ethical sanctity of law. Mortgage and sale of *patta* land is the most common form of land alienation, especially in the scheduled areas. Driven by poverty and indebtedness, people have no choice but to sell or mortgage the land they have received from the government. These poor families get a very meager amount in exchange for their land. Such mortgage takes place both within and outside the tribal community.

LAND ALIENATION (ALL CATEGORIES) IN STUDY VILLAGES OF KORAPUT DISTRICT

Village-wise number of transfers of land through different methods and extent of area transferred is mentioned in Appendix-VI. Out of all types of land transfers, the highest number (78) of transfers through mortgage was reported followed by transfer through sale (55). In case of other types of transfers including mortgage to money-lenders, oral transfers, lease, encroachment (forcible occupation), public purpose, normal transfer, land acquisition and cheating etc. it was found that lands of 82 households were transferred.

Land Acquisition

The village-wise data indicates that only one person in Dhaudapadar lost 18 acres of land due to land acquisition. Land acquisition in this village was done 10 years before. In Machhput as many as 13 HHs lost 119.82 acres land in land acquisition. While interacting with the villagers during FGD, it came to notice that they were allotted for land and paid some amount of cash compensation. But they had not shifted to the allotted site on the ground that the area was not suitable for settlement and they could not have adjusted there. The lands were acquired for Kolab project. The villagers of Nighamaniguda adduced the same reason where 21 people lost 110.09 acres of land due to land acquisition for upper Kolab project. In Sriramput village, 5 HHs surrendered their land due to acquisition. Altogether 42 HHs fell victim to land acquisition.

Sale

The maximum number of transfers (22 HHs) in terms of sale had been reported from village Mania followed by village Panchada (i.e. 13 HHs). No transfer was found in the form of sale in Machhput. In other villages, about two to five transfers had been noticed.

Mortgage

It was found that maximum transfers had taken place in the form of mortgage. Mukhibedei village had the highest number

(29) of mortgages covering an extent of 60.75 acres. No mortgage case was however, found in Devighat and Bhitaragada villages.

Cheating

It was alleged by some HHs of Srirampur village that their lands were transferred through cheating. In Mukhibedei and Mania villages such cases of cheating (one each) were also reported. In the remaining four cases, oral transfer was reported, two from village Dhaudapadar and one each from Nigamaniguda and Mukhibedai villages. For instance, one cheating case reported by the victims of Srirampur village where one official took away the lands by manipulating caste certificate without leaving any authentic evidence. The officer threatened him not to take any step against him.

Lease case

Except one lease case in Mania village, nowhere lease cases have been reported.

Forcible occupation

About 60 acres of lands are forcibly occupied (encroached) by both tribals and non-tribals. Out of these cases, five cases were reported from village Panchada followed by Mania (4 no.), and Srirampur (3 no.), Devighat and Nighamaniguda villages with two each.

Land Alienation of ST HHs in Study Villages of Koraput and Rayagada Districts

In regard to the transfer cases of ST households of Koraput district exclusively, it was found that maximum transfer has taken place through mortgage (51 Nos.) and by sale (21 Nos.). Forty-nine households had their lands transferred through other methods like oral transfer, encroachment/forcible occupation, public purpose, land acquisition, cheating and other types like *benami* transfer etc. (see Appendix-VII). The field survey on method of transfer of land of ST households in the study villages of Rayagada district, indicate that a total of 78 transfers had taken place, out of which 22 households transferred land by sale, 41 households transferred their land by mortgage and the rest 15 households transferred their lands through other methods as shown in Table 6.8.

Table 6.8: Land alienation of ST households

Name of the District	Method of transfer			
	Sale	Mortgage	Other transfers	Total ST transfers
Koraput	21(17.35)	51(42.15)	49(40.50)	121(100.0)
Rayagada	22(28.21)	41(52.56)	15(19.23)	78(100.0)
Total	43(21.61)	92(46.23)	64 (32.16)	199(100.0)

Note: Figures in the bracket indicate percentage to the total.
Source: Field survey 2008-2010.

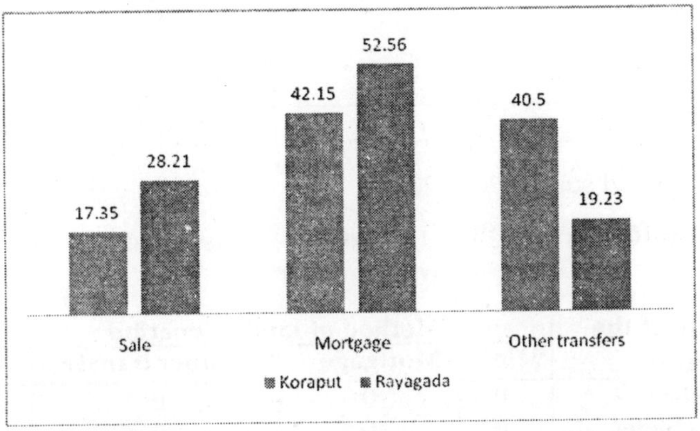

Figure 6.2: Land alienation of ST households in Koraput and Rayagada district.

Since majority of land transfers were found in tribal households, the study focused on village-wise transfers of ST lands in Koraput and Rayagada districts. In Koraput district (Table 6.9), maximum transfers were noticed in Machhput, Nigamaniguda and Mukhibedai villages. Similarly, in Rayagada district (Table 6.10), maximum transfers were found in Bhatpur, Matikona and Tadama villages.

Table 6.9: Land alienation of ST households in the study villages of Koraput district

Name of the villages	Method of land alienation			
	Sale	Mortgage	Other transfers	Total
Devighat	3	-	1	4
Dhaudapadar	2	6	2	10
Machhput	-	3	14	17
Nighamaniguda	2	1	20	23
Bhitaragada	-	-	-	-
Baraguda	-	-	-	-
Panchada	3	3	3	9
Jhodipadar	-	-	-	-
Mukhibedei	2	25	2	29
Sriramput	2	2	7	11
Mania	5	2	-	7
Aligan	2	9	-	11
Total	21	51	49	121

Source: Field survey 2008-2009.

Table 6.10: Land alienation of ST households in the study villages of Rayagada district

Name of the villages	Method of land alienation			
	Sale	Mortgage	Other transfers	Total
Garadi	0	0	0	0
Chanchadaguda	0	0	0	0
Bhatapur	4	4	2	10
Majhihalma	0	0	1	1
Gujalpadu	1	3	0	4
K. Dengasargi	0	1	1	2
Gunakhal	1	2	2	5
Bariguda	0	1	1	2
Matikona	12	8	3	23
Laxmipur	3	0	0	3
Nathama	0	1	4	5
Tadama	1	21	1	23
Total	22	41	15	78

Source: Field survey 2009-2010.

Causes of Land Alienation

Agriculture is the predominant occupation and major source of livelihood among the tribals. This concept proved to be wrong in the present study. In the study area, the main source of livelihood was dependence on daily wage labour. This may be due to either the households possess insufficient quantity of land as a result of land alienation or they are incapable of cultivating the available land with them due to financial constraints. Alienation, as mentioned above, has been in the form of sale, mortgage, encroachment (forcible occupation) and other types of transfer like *benami* transfer etc. Lands of tribals are transferred out of compulsion in some cases and in some other cases out of ignorance. The compulsions for land transfer include household expenditure, house construction, repayment of old debts, socio-religious obligation, agricultural improvement, land development, land purchase, purchase of livestock, medical treatment, alcoholism, drought, inability to cultivate and due to government purchasing/acquiring lands for development projects. To meet their requirements, the tribals started to transfer their movable properties initially and finally transferred immovable property including land by way of mortgage. In absence of regular source of income to repay old debts and meet chain of social and religious obligations and the fear of dying indebted and constant reminders by the money-lenders, the tribal borrower finally parted with his land to avoid unwanted embarrassments. Some of the cases are recorded during field study (mentioned in Appendix-VIII). Although the tribal borrower has the option of either losing his land or taking recourse to existing laws, he normally prefers the former by taking the moral responsibility of non-repayment of debt himself. But this trend gradually has undergone changes after the restoration of lost tribal lands by competent authorities under the law.

The present study tried to enquire about the personal reasons for land alienation in the form of sale and mortgage. Table 6.11 (a) reveals the reasons for sale of land in the study villages of Koraput district. Out of total 55 land alienation cases due to sale, 13 HHs reported that due to their need for purchasing bullocks, they sold some portion of their land. Twelve households attributed the reason to meet household expenses and 10 HHs to meet cost for medical treatment.

Table 6.11 (a): Land Alienation through sale in Koraput district

Name of the village	Causes of land alienation													
	1	**2**	**3**	**4**	**5**	**6**	**7**	**8**	**9**	**10**	**11**	**12**	**13**	**Total**
Devighat	0	0	0	1	0	0	0	0	1	0	0	0	0	2
Dhaudapadar	1	1	0	0	0	0	0	0	1	0	0	0	0	3
Machhput	1	0	0	0	0	0	0	0	0	0	0	0	0	1
Nighamaniguda	3	0	0	0	0	0	0	0	0	0	0	0	0	3
Bhitaragada	0	0	0	0	0	1	0	0	0	0	0	0	0	1
Panchada	3	0	2	1	0	0	0	0	2	1	0	1	0	10
Mukhibedei	1	0	0	0	0	0	0	0	2	0	0	0	0	3
Sriramput	0	0	0	0	0	0	1	0	1	0	0	0	0	2
Mania	3	1	0	0	1	11	0	2	3	0	0	0	3	24
Aligan	0	1	0	0	1	1	0	0	0	0	1	0	2	6
Total	12	3	2	2	2	13	1	2	10	1	1	1	5	55

Source: Field survey 2008-2009.

Table 6.11 (b) reveals the reasons for sale of land in the study villages of Rayagada district. It is found that out of 22 land alienation cases through sale, 17 households reported that sale was to effected meet the household expenses while 2 of the households attributed the reason to repayment of debt. Social and religious reasons, medical treatment and inability to cultivate the land were the other reasons cited by the households in each of the above cases. There was no case of land alienation through sale in six out of twelve study villages namely, Garadi, Chancharguda, Majhihalama, K.Dengasargi, Bariguda and Nathama.

Table 6.11 (b): Land alienation through sale in Rayagada district

Name of the village	Causes of land alienation													
	1	**2**	**3**	**4**	**5**	**6**	**7**	**8**	**9**	**10**	**11**	**12**	**13**	**Total**
Garadi	0	0	0	0	0	0	0	0	0	0	0	0	0	0
Chanchadaguda	0	0	0	0	0	0	0	0	0	0	0	0	0	0
Bhatapur	2	0	0	1	0	0	0	0	0	0	0	1	0	4

Name of the village	Causes of land alienation													
	1	2	3	4	5	6	7	8	9	10	11	12	13	Total
Majhihalma	0	0	0	0	0	0	0	0	0	0	0	0	0	0
Gujalpadu	1	0	0	0	0	0	0	0	0	0	0	0	0	1
K. Dengasargi	0	0	0	0	0	0	0	0	0	0	0	0	0	0
Gunakhal	1	0	0	0	0	0	0	0	0	0	0	0	0	1
Bariguda	0	0	0	0	0	0	0	0	0	0	0	0	0	0
Matikona	10	0	2	0	0	0	0	0	0	0	0	0	0	12
Laxmipur	2	0	0	0	0	0	0	0	1	0	0	0	0	3
Nathama	0	0	0	0	0	0	0	0	0	0	0	0	0	0
Tadama	1	0	0	0	0	0	0	0	0	0	0	0	0	1
Total	17	0	2	1	0	0	0	0	1	0	0	1	0	22

Source: Field survey 2009-2010.

Note: 1. Household expenditure; 2. House Construction and Repairing; 3. Repayment of Debt; 4. Social and Religious Functions; 5. Agricultural Improvement; 6. Bullock Purchase; 7. Purchase of raw-materials; 8. Court case; 9. Medical Treatment; 10. Alcohol/Drugs; 11. Natural calamities 12. Unable to cultivate and 13. Crop Failure.

Note is for Both Table 6.11 (a) and 6.11 (b)

With regard to the reason for mortgaging land in the study villages of Koraput district, it was found that 21 HHs have mortgaged their land to meet the cost of medical treatment, 15 households for purchase of bullock, 11 HHs for meeting household expenses and the rest for other reasons as indicated in Table 6.12 (a).

Table 6.12 (a) : Land alienation through mortgage in Koraput district

Name of the village	Causes of land alienation through mortgage										
	1	2	3	4	5	6	7	8	9	10	Total
Devighat	0	0	0	0	0	1	1	0	0	1	3
Dhaudapadar	1	0	0	0	0	2	1	1	0	0	5
Machhput	2	0	0	0	0	0	1	0	0	0	3

Name of the village	Causes of land alienation through mortgage										
	1	2	3	4	5	6	7	8	9	10	Total
Nighamaniguda	1	0	0	0	0	1	1	0	1	0	4
Panchada	1	1	1	2	0	0	2	0	0	3	10
Mukhibedei	4	2	1	3	1	3	11	0	0	3	28
Sriramput	0	0	0	0	0	1	1	0	0	2	4
Mania	1	2	0	0	0	4	2	0	0	3	12
Aligan	1	0	2	0	1	3	1	0	0	1	9
Total	11	5	4	5	2	15	21	1	1	13	78

Source: Field survey 2008-2009.

In the study villages of Rayagada district, it may be seen from Table 6.12(b) that a total of 41 transfers had taken place through mortgage. Twenty households reported the reason of transfer to meet the expenses for agricultural improvement, 8 households to meet household expenditure, 5 households for house construction and repairing, 3 debt repayment and the rest mortgaged their land for other reasons as shown in Table 6.12 (b).

Table 6.12 (b): Land alienation through mortgage in Rayagada district

Name of the village	Causes of land alienation through mortgage										
	1	2	3	4	5	6	7	8	9	10	Total
Garadi	0	0	0	0	0	0	0	0	0	0	0
Chanchadaguda	0	0	0	0	0	0	0	0	0	0	0
Bhatapur	2	0	2	0	0	0	0	0	0	0	4
Majhihalma	0	0	0	0	0	0	0	0	0	0	0
Gujalpadu	2	0	0	0	0	0	0	0	0	1	3
K. Dengasargi	0	0	0	0	0	1	0	0	0	0	1
Gunakhal	1	0	0	1	0	0	0	0	0	0	2
Bariguda	0	1	0	0	0	0	0	0	0	0	1
Matikona	3	2	1	0	0	0	0	1	1	0	8

Name of the village	Causes of land alienation through mortgage										
	1	**2**	**3**	**4**	**5**	**6**	**7**	**8**	**9**	**10**	**Total**
Laxmipur	0	0	0	0	0	0	0	0	0	0	0
Nathama	0	1	0	0	0	0	0	0	0	0	1
Tadama	0	1	0	0	20	0	0	0	0	0	21
Total	8	5	3	1	20	1	0	1	1	1	41

Source: Field Survey 2009-2010.

Note: 1. Household expenditure; 2. House Construction and Repairing; 3. Repayment of Debt; 4. Social and Religious Functions; 5. Agricultural Improvement; 6. Bullock Purchase; 7. Medical Treatment; 8. Alcohol/Drugs; 9. Unable to cultivate; and 10. Other Causes.

Note is for Both Table 6.9(a) and 6.9(b).

Restoration of Alienated Lands

Land Alienation over the years has been the single biggest cause of acute backwardness of tribal economy. A sizable portion of the tribal population is economically indigent and has been reduced to the status of landless agricultural labourers as a result of gradual alienation of lands from tribals to non-tribals including a few rich tribals. The tribals lost their sense of security and are unable to repay the loans taken from the money-lenders. Quite a few of them surrendered their liberty of working, and offered them as bonded labourers for generations (Mahapatra, 1987). Following the occurrence of several tribal revolts, the government in the past had passed several legislations/regulations to control alienation of tribal lands and restore them to the owners.

The present study revealed that the tribals are hardly aware of the constitutional rights of their entitlements. Majority of the tribals in the study area do not know that their alienated land to the non-tribals can be restored. Even they sometimes and in some cases feel guilt for filing cases for restoration. In some cases, as it has been observed in Rayagada district, the parties who received notice for attending courts for the hearing of restoration cases of illegally transferred lands, simply ignored such notices for the reason that they did not have money to attend the courts of collector/sub-collector. During the course of interaction, they of course showed their interest for restoration. In Koraput district, during the process of reality check, it was found that the tribal lands which were restored as per government record were not in their actual physical possession and even they were not informed of such restoration. One of the former Tahasildars of Koraput stated that the restored land is again re-occupied by the non-tribals and in that case, government can not help as it is difficult to identify such cases.

State Protective Legislations for Restoration of Alienated Lands

In the early part of the 20th century, in large areas of tribal regions, land alienation took place without any documentation. In 1917, the British government passed the Agency Tracts Interest and Land Transfer Act (Act 1 of 1917) wherein regulatory provisions were made regarding transfer of land by tribals to non-tribals. This Act was comparatively more helpful in plain and accessible areas than backward areas. It is only in the post-independence period that a series of legislative measures were taken to prevent transfer of tribal lands to non-tribals. In the post-independence period, Orissa has undertaken a number of legislative measures to prevent economic exploitation in the tribal regions of the state. Similarly, legislation has also been passed to abolish debt bondage, provide for debt relief and regulate money-lending in tribal areas. In this regard the (i) Orissa Debt Bondage Abolition Regulation, 1948 (Regulation 1/48), (ii) Orissa Scheduled Areas Transfer of Immovable Property (by Scheduled Tribes) Regulation, 1956 (II/1956), (iii) Orissa Scheduled Areas Transfer of Immovable Property (by Scheduled Tribes) Amendment Act, 1975, (iv) Orissa (Scheduled Areas) Money Lending Regulation, 1967, and (v) Orissa (Scheduled Areas) Debt Relief Regulation, 1967 have been enacted.

The Orissa Scheduled Areas Transfer of Immovable Property (OSATIP), 1956 Regulation (2) restricted land transfer from tribals to non-tribals and provides for restoration of illegally alienated lands to the tribals. Section 23 and 23 (A) of Orissa Land Reforms (OLR) Act, 1960 provide for restoration of illegally alienated land to the tribal owners. The Revenue Officer, on his own information or on the application of any interested person may declare such a transaction invalid and restore such property to the owner or his heirs. Sec.23 (A) of the OLR Act provides that if a non-tribal is found to be in unauthorized occupation of the land of a tribal person by way of trespass or transfer in contravention of the laws, the Revenue Officer may order eviction and shall cause restoration of such property to the tribal owner or his heirs. Under Section 22 of OLR Act, 1960, any transfer of holding by a *raiyat* belonging to STs shall be void except in case it is made in favour of a person belonging to STs or by SCs to SCs and if transferred by STs/SCs to non-STs/SCs without permission of the competent authority or Revenue Officer in other than scheduled areas. Under sections 23 and 23 (A) of OLR Act, the Revenue Officer or competent authority shall act *suo motu* or on the application of any person interested for causing restoration of the property to the transferor or his heir, or if it has been transferred without prior written permission of competent authority. Section 3 (2), 3(A) and 7(A) of Regulation 2 of 1956 provide for restoration of any land transferred against the law. Sec 3 (2) empowers the competent authority to order ejectment of the non-tribal person and restore the land to the tribal owner or his heirs. In case, restoration is not reasonably practicable, the authority may settle the property with another member of scheduled tribe or in the absence of any such member, with any other in accordance with the provisions contained in the Orissa Government Land Settlement Act of 1962. Section 7(A) states that despite an eviction order, if a non-tribal person is found to be in reoccupation of the immovable property of a tribal person, the former shall be punishable with rigorous imprisonment or fine. Table 6.13(a) indicates the data on cases instituted. In Koraput district, out of total 391 cases instituted, altogether 284 restoration cases were instituted on application and 107 on *suo mottu* action in which 182 cases were restored.

Table 6.13 (a): Cases instituted in Koraput district

Section	Year					On application	Suo Motu
	2005	2006	2007	2008	Total		
3(2)	98	26	5	11	140	86	54
3(4)	64	14	32	41	151	115	36
3(4) (1)	-	-	6	17	23	6	17
2(2)	17	20	11	5	53	53	-
23(4)	14	6	3	1	24	24	-
Total	193	66	57	75	391	284	107

Source: Sub-collector office, Koraput.

Table 6.13(b) indicates the position of disposal of cases under regulation of 1956 for the year 2007-08 in Rayagada district which indicates that 2848 number of total cases were instituted for disposal out of which 422 were disposed off and 83 persons were benefited thereby.

KEY FINDINGS

1. Among all types of transfer of land in Koraput district, maximum transfer was found through mortgage (78 nos.) followed by sale (55 nos.) and 82 through other methods. In respect of scheduled tribe households, the same trend was noticed i.e. maximum through mortgage (51 nos.) followed by sale (21 nos.) and 49 numbers through other methods.
   In the study villages of Rayagada district, altogether 78 cases of transfer had taken place, out of which 22 households transferred land through sale, 41 households transferred by mortgage and rest 15 households through other methods of transfers like oral transfer, encroachment/forcible occupation, public purpose, land acquisition, cheating and *benami* transfers etc. In both the districts, transfer was effected mainly through mortgage.
2. In Koraput district, out of total 55-land alienation cases through sale, 13 HHs reported that they alienated land

Table 6.13(b): Disposal of cases under regulation 2 of 1956 for the year 2007-08 of Rayagada district

Name of the cases	No. of cases pending as on 1.4.07	Cases instituted from 1.4 07 to 31.03.08	Total cases for disposal	No. of cases disposed off up to 31.3 08	Balance	% of disposal	No. of cases in which orders have been passed		No. of persons bene-fited	Extent of land resto-red	Balance area for resto-ration	
							No.	Area			No.	Area
1	2	3	4	5	6	7	8	9	10	11	12	13
Sub section-3-a (i)	21	0	21	0	21	0	0	0	0	0	0	0
U/S 3(2)	0	1	1	0	1	0		0	0	0	0	0
U/S 3 (A)	21	5	26	13	13	50	13	26.42	5	12.09	105	32.853
U/S 3 (B)	2717	83	2800	409	2391	14.61	409	544.319	78	151.786	0	0
TOTAL	2759	89	2848	422	2426	14.82	422	570.739	83	163.876	105	32.853

Source: Annual administrative report, Rayagada district, 2007-2008.

for purchasing bullocks. Twelve households alienated land for meeting household expenses and 10 HHs to meet medical treatment. In Rayagada district, out of 22 land alienation cases through sale, seventeen alienated to meet household expenses while in 2 households, alienation was done for repayment of debt. In other cases, alienation had taken place due to social and religious reasons, medical treatment and inability to cultivate the land. However, no case of land alienation through sale was reported in six out of twelve studied villages. These villages are Garadi, Chancharguda, Majhihalama, K.Dengasargi, Bariguda and Nathama.

3. The reason for mortgaging land in the study villages of Koraput district in respect of, 21 HHs was to meet requirement for medical treatment, 15 households for purchase of bullock, 11 HHs for meeting household expenses and the rest for other reasons. In case of Rayagada district, out of 41 transfers through mortgage, 20 households mortgaged land to meet the expenses for agricultural improvement, 8 households to meet household expenditure, 5 households for debt repayment, 3 households to meet expenses in social and religious functions and the rest mortgaged for other reasons.
4. In Koraput district, out of total 391 cases of restoration of lands, 284 cases were instituted on applications filed and 107 by *suo motu* initiation by competent authorities. Out of these, in 182 cases, lands were restored back during 2005-2008. In Rayagada district, out of total 568 cases including regulation 2 of 1956 and OLR, 1960 cases, in 238 cases lands were restored back during the period 2005-2008.
5. The latest amendments to 1956 Regulation 2 have invited several adverse situations. Under the regulation, ban on transfer of lands of ST persons in section 3 has caused serious hardship. This has encouraged formal (concealed) transfer of land through unregistered mortgage/sale involving actual possession with tacit understanding of both the parties.

This is otherwise called *benami* transaction undetected by enforcing authorities.

6. Some of the terms like 'land', 'tribal land', 'fraud' and 'defraud' referred to in section 3-B of Regulation 2 of 1956 needs further elaboration to wipeout some of ingruities in the law. For example, section 3-A envisages eviction of persons in unauthorized occupation of property of a member of ST by any person for restoration but section 3-B provides for reversion/restoration of lands of STs which are found to have been transferred by fraud. While under section 3-A appeal and revision to the next higher designated authority is provided, under section 3-B provision of appeal/revision is conspicuously absent. This amounts to denial of right/opportunity to the aggrieved to seek redress against injustice. This omission needs correction.
7. Under section 3-B person who is in possession of agricultural land alone belonging to a ST member is required to notify to the sub-collector in the prescribed manner as to now he came in possession of the land but it does not make any distinction between ST & non-ST. Only those non-ST persons who acquire lands belonging to STs on the authority or permission only came forward to notify the manner of occupation and acquisition but there are genuine cases covered under due permission which are also being probed. This procedure, therefore, needs further clarification.

7

Impact of Act/Laws—Gaps and Suggestions

In spite of several protective legislations, the interest of tribals continues to be unprotected. The present study tries to assess the impact of the legislations and locate the gaps in the laws/ Acts. The suggestions emerged from the opinions of various key stake-holders and study results are also focused for considering those as policy recommendations.

Government of India has adopted several programmes and plans to protect the rights and security of tribals. But there have been violations of their fundamental rights to live as humans with dignity and equality. During the course of history, at a particular period of time, tribal villages and settlements noticed to have disappeared and tribals were deprived of their livelihood resources. Despite legal protection, land alienation has deprived many tribals, *dalits* and other marginalized people from their rights and entitlements. For example, under the Land Acquisition Act of 1894 as amended in 1984 and the Coal Bearing Areas Act of 1957, government is competent to acquire land for public purpose on payment of compensation for strategic and infrastructure development projects. But there is a big question mark whether such compensation is adequate or not. As most of the development projects including mining and hydropower are located in tribal areas, the tribals are the main and major affected persons due to land acquisition. In consequence, many tribals have become landless. In spite of all compensatory measures adopted by government, the socio-economic and cultural life of tribal people hardly get settled to their original living conditions. As a result, the so-called "public purpose" for tribal people gets defeated. Although government officers act as agents of government in acquiring tribal lands, practically they are unable to deliver and pay appropriate compensation (Patnaik, 2009) and such type of transfer invariably, favours the government and certainly not the tribals. Although, the Land Acquisition Act had undergone minor amendments in 1914, 1938 and substantially in 1984 to streamline the process of Acquisition, different study results have shown that compensation paid for lands acquired is not enough in comparison to the actual market price of land. Besides, delay in payment of compensation is also invariably noticed.

Similarly, under sections 23 and 23A of OLR Act, the Revenue Officer or competent authority shall act *suo motu* or on the application of any person interested for causing restoration of the property to the owner or his heir if it has been transferred without prior written permission of competent authority. But implementation of this Act has been ineffective in as much as that in many cases, although restoration has been shown in pen and paper, physical possession of such land had not been restored. (Patnaik, 2009) Likewise illegal transfers from tribal to non-tribal under OSATIP Regulation of 1956 [Regulation (2)] encouraged fraudulent practices which compelled the state government to go for further amendment in 2002. As required under the Panchayats (Extension to Scheduled Areas) (PESA) Act, 1996 the state government has amended (OSATIP, 1956) in providing for a stringent restriction of complete ban on transfer of lands of scheduled tribes to non-tribals. This provision in the 2002 amendment was a landmark confirmatory legislation, which, in the meantime, has raised several controversial questions. The state government is reportedly on an exercise in bringing further amendment to provide exceptional space in this law to allow scheduled tribes to transfer land on certain exceptional situations.

In order to assess the impact of the Acts/Laws [Special focus on 2002 Amendment Act of OSATIP, 1956, Reg (2)], role of Gram Sabha under PESA Act, acquisition of land and status of possession of lease lands in terms of gaps and benefits in the implementation process in the study areas, the following tools were adopted:

1. Interview with Sarapanches and interaction with local people during conduct of FGDs.
2. Informal interviews with key informants like government officials, elected representatives, activists, NGOs, advocates and academicians.
3. Interaction with the key persons of the villages.
4. Reality check/cross-verification—The purpose was to check whether information with respect to compensation for land acquisition, issue of *patta* and possession of lease land and restoration of tribal lands has matched the HH information or there exists any gaps.

The first tool was for the assessment of role of Gram Sabha in tribal land issues under PESA Act. The second and third tools were adopted to assess the impact of Orissa Scheduled Areas Transfer of Immovable Properties (OSATIP) of 1956 Regulation (2) followed by 2002 amendment which has been implemented to protect tribal land restricting total transfer of land from tribal to non-tribal. The fourth tool was to examine the implementation of Land Acquisition Act, strategy for distributing lease *patta* and restoration under 2002 amendment of OSATIP, (Regulation 2/56) through cross-verification (reality check) in the study villages.

A. ASSESSMENT OF ROLE OF GRAM SABHA IN LAND RELATED ISSUES UNDER PESA

Role of Gram Sabha under PESA Act

The provisions concerning Panchayats in the 73rd Constitutional Amendment Act were extended to the Scheduled Areas through an Act of Parliament. The provision of Panchayat (Extension to the Scheduled Areas) Act, 1996 came into force with effect from 24.12.1996. Subsequently, many states having scheduled areas had to come out with their respective confirmatory Acts before 24.12.1997 or before the expiration of one year from the date on which the central Act received the assent of the President. Accordingly, the Orissa Panchayat Acts were amended in confirmatory with the aforesaid Central Act and got the assent of the Governor on 21st December, 1997 and were made applicable to the tribal districts of the state with immediate effect. The Central Act 40 of 1996 has also clearly defined the constitution of a Gram Sabha in the scheduled areas. Gram Sabha will be competent to safeguard and preserve the traditions and customs of the people, their cultural identity, community resources and customary mode of conflict resolution. Under the mandatory powers, the Gram Sabha is empowered to approve the plans, programmes and projects for social and economic development before the Panchayat takes them up; to identify or select beneficiaries of poverty alleviation and other such programmes; and to issue utilization certificate to the Panchayat for the amount utilized by the later for the development of the village.

Under the provisions of the Act, the Gram Sabha needs to be consulted with regard to the acquisition of land for developmental projects and before resettling or rehabilitating the affected persons by such projects in the scheduled areas. This is in conformity with the traditional pattern of land ownership in many tribal areas where land is a common property and owned by the village as a whole rather than by any single individual or individuals. The Act also makes the provision that necessitates the Gram Sabha's recommendation for the grant of a prospecting license or mining lease for minor minerals in the scheduled areas as well as for that of concession for the exploitation of minor minerals by auction if the above sites are under the jurisdiction of the village.

The Act further specifies that the respective state legislature must ensure that the Gram Sabha or the Panchayats at appropriate level are endowed with the following powers (i) to enforce prohibitions or to regulate or restrict the sale and consumption of an intoxicant, (ii) ownership of minor forest produce, (iii) to manage village markets, (iv) to exercise control over money-lending to the scheduled tribes, and (v) to exercise control over institutions and functionaries in the social sectors and local plans and resources including tribal sub-plans. The state government will have to entrust the minor water bodies to the Panchayats for arrangement. This Act, therefore, extends enormous powers to the Gram Sabha to self-regulate and for self-governance. Under Section 4 (m) (iii) of the Central Act No. 40, the Gram Sabha is further vested with the power to prevent alienation of land in the scheduled areas and to take appropriate action to restore the unlawfully alienated land of scheduled tribes.

To justify this in reality, questions pertaining to conduct of Gram Sabha, agenda of Gram Sabha, number of participation, etc. were queried during FGDs and informal interviews with Sarpanches. It is revealed that there is regular conduct of Gram Sabha sittings 2 times in a year and somewhere 4 times depending upon the need. Participation of women in Gram Sabha is encouraging. It was noticed that the members participate proactively on issues like IAY, drinking water, PDS and other developmental works. Land related issues are not discussed in Gram Sabhas except in case of Gram Sabha of Devighat Panchayat in Koraput block. Interestingly, the Sarpanch or the villagers have hardly any idea about the role of Gram Sabha in land related issues.

During the course of interaction with the Sarpanches in the study GPs, it came to notice that land issues had never been included as a subject in the Gram Sabha agenda. In exceptional cases only, the issues are initiated by Sarpanches occasionally. It was noticed that Mr. Bibhuti Bhusan Nayak, Sarpanch of Devighat GP of Koraput arranged to file 203 lease cases and encroachment cases of the villages of the GP for getting *patta*. Similarly, Marli Drulia, Sarpanch of Rajput GP played an active role in land acquisition for water supply and road construction works under PMGSY. For this, Pallisabha was conducted in Rajput village. Prior to construction of PMGSY road from Pitagada to Patraput, he called a Pallisabha for taking decision in providing work to the people who would be expected to loose their land in the process of acquisition. The Sarpanch assured to ensure for providing compensation in the form of work and IAY houses. Ms. Geeta Muduli, Sarapanch of Padmapur, reported that Gram Sabha is conducted regularly and the issues in Gram Sabha are raised mostly with regard to developmental works only. But no issues relating to land alienation ever came up for discussion. This might be due to (i) either people have no idea that these issues should be discussed in the Gram Sabha meetings, (ii) Sarpanch is also not initiating these issues to be raised from his/her side in Gram Sabha. After the trainings were conducted by ISS, Sarpanches now seem to be interested to take up land issues in Gram Sabha meetings henceforth. The Sarapanch of Dumuriguda GP, Rayagada Block reported that Gram Sabha is not playing any role in settling land related disputes. Participation of women in the Gram Sabha is less than the male participation.

B. ASSESSMENT OF 2002 AMENDMENT OF OSATIP, 1956, REGULATION 2

Gaps

The amended regulation of 2002 has raised a number of controversial issues which emerged during interaction with elected representatives, activists, NGOs, advocates, academicians and the key persons of the villages. These are briefly indicated below:

- The tribal landholder, who has even 20 acres of land, cannot now transfer land to a non-scheduled tribal person as per the amended provision of 2002. If he wants to sell the land to a tribal, it is very difficult to get such a party. Now the problem is that whenever any emergency occurs in the family and if the family has no money to meet expenses for such emergency situations, possession of such land as an asset, becomes redundant because of the present legal restrictions. Even persons could die without treatment for not having enough cash with them to meet such emergencies. Hence having land or no land makes any difference to them in the circumstances.
- In certain cases, a non-tribal who wants to take the land from a tribal, is trying to manipulate caste of the tribal person. In Sriramput village of Rajput Gram Panchayat of Semiliguda block, in Koraput district, the inhabitants

belong to Katia Adivasi, a type of tribal having their surname "Rana". There are persons having the surname "Rana" who belong to OBC type also. Taking advantage of this anomalous position, a particular non-tribal person got some lands illegally transferred in his name.

- Non-tribals who acquire tribal lands through sale/ mortgage are facing a lot of problems. As pointed out by Mr. Bibhuti Bhusan Nayak, Sarpanch of Devighat GP that a non-tribal person of his Panchayat acquired a piece of land from a tribal on payment of some amount of money. The non-tribal was yet to start cultivation. In the meantime, he got notice for restoring back the land of the tribal. In this case, the non-tribal was looser in both ways. In one way, he did not get any benefit from the land and on the other hand, he lost his money in paying the tribal for keeping the land with him through mortgage.
- Land restored to tribals has again been alienated to non-tribals and in some cases tribals refused to cultivate the restored land for various reasons. Some of the tribals feel in-secured and are afraid of reprisals from influential non-tribals as reported during interaction with some key persons of the study area.
- Mr. Bhagat Jani of RI, Kakiriguma of Koraput district explained about some of the realities in the implementation process and suggested the following:
    (a) The transfers of land through sale are mostly verbal transfers. As a result, such transactions do not come to the knowledge of authorities.
    (b) Even if land has been transferred from ST to general category of people with all documents, it is very difficult to get the land back.
    (c) Since years, the settlement of land has not been updated in absence of which many problems could not be solved.
    (d) Lack of awareness about the Acts also creates problems.
    (e) Non-tribals dominate tribals. So tribals are suppressed even though they are aware of certain things.

(f) In some cases where tribal land is restored back by non-tribals, then that land again is being snatched away by the non-tribals.

- Dr. P.C. Mohapatra, Director COATS, Koraput indicated that 1956, Regulation (2) was not working effectively because of lacuna in the law, ignorance of the tribals and complicated legal provisions. Rather *benami* transactions take place.
- Advocate, Sri B.S. Nayak, Special Public Prosecutor and Vice-Chairperson, Rayagada Municipality, having legal expertise in land related cases in tribal areas mentioned that government distributes lease *patta* but sometimes it creates problems. For instance, government distributes *patta* without showing the land physically which means land *patta* only was issued but land had not been shown nor demarcated. So the *patta*-holders do not have any idea of their own allotted land. If some one occupies and cultivates land without knowing possession of land, in that case, a man who starts cultivating land of another, disputes arise between the parties. Mr. Naik pointed out that this is a serious gap in distribution of leased *patta* land.
- Another problem pointed out by him was with regard to partition cases. While applying for partition among the entitled landholders, without any reluctance, they sign the papers but are not attending to depose their statements before Tahasildar. As a result, the cases are pending. Although the *patta* simply remains in the name of the person but the *patta* land is actually cultivated by someone else. This is due to lack of awareness and illiteracy only. Therefore, there is need for more training and awareness programmes among the illiterate tribals.

C. REALITY CHECK

Reality check is an act of verifying data as per government record in the field. Using this tool, the gaps/impact of the regulation can be identified. It can help government to undertake reform measures at the policy level. Keeping this in view, the present study tried to verify some government records/information in matters relating to land acquisition, issue of *patta* of lease lands and restoration of tribal land in the study villages adopting the following methodology:

The present study was confined to 24 villages of two districts. The related data of those villages were collected from the concerned departmental offices of the Government. A format was prepared for separate subjects viz. (i) land acquisition, (ii) distribution of lease land, (iii) restoration of land. The enlisted HHs under the above subjects of study villages were checked as per the format. The purpose was to check whether government information matched with the HH information collected and if there exists any gap between the two.

Land Acquisition

While verifying land acquisition position, it was found that only three villages namely Machhput, Mania and Bhitaragada of Koraput district were victims of land acquisition due to Upper Kolab and NALCO Projects respectively. No acquisition cases were found in the study villages of Rayagada district. Hence, no verification was done in this district. The position of land acquisition was therefore, verified in the affected study villages of Koraput district only. The position of Koraput district is reflected in Table 7.1.

Table 7.1: Land acquisition in the study villages of Koraput district

Name of the village	Name of the project	Year of acquisition	No. of the persons checked	Land acquired (in acres)	Compensation (In cash/job)
Mania	NALCO	1982	10	1-3 acres	Rs. 3000-7000 No job provided
Bhitaragada	NALCO	1982	5	30-70 cents	Rs. 900-1300
Machhput	Upper Kolab	1980	4	0.23-2.59 acres	Rs. 478-6639

Source: (i) Land Acquisition Section, Collectorate, Koraput.
(ii) Land Acquisition and Resettlement: Upper Kolab Project, Koraput.

In respect of land acquisition, 10 HHs from Mania village, 5 HHs from Bhitaragada and 4 from Machhput villages were checked. The information collected from the government offices was almost matching with the HH information collected from Mania and Bhitaragada villages. About payment of compensation, it was found that cash amount paid was matching with official figures but nobody was provided with job as committed at the time of land acquisition. In case of Machhput, no HH was found in government records in respect of acquisition. As per the version of the local people neither any person was affected nor anyone left this village, nor any such person is living. Hence, in Machhput village, the data could not be checked.

Distribution of Lease Land

There are a large number of families who have been allotted land but have not got *patta* implying thereby the inherent lacunae in the implementation of land reforms during the last three decades. Sometimes, the beneficiaries being ignorant about official procedures hesitate to discuss such matters with the concerned Revenue Inspector. Sometimes, their financial conditions do not allow them to probe into matters in depth. In many villages, the land allotted to the legitimate beneficiaries was either unproductive or uncultivable. Negligible or no effort has gone into developing the land to make it cultivable. Moreover, the allotted land is usually located at a far-away place from the village and sometimes adjacent villagers claim rights over the reallocated land. Physical land demarcation is also a major issue and a majority of beneficiaries are unable to identify their plots. As a result, they cultivate land which does not belong to them and are, therefore, called encroachers. There are many such examples, especially in districts like Gajapati, Kondhamal and undivided Koraput (Status Report, UNDP, 2008). Taking all these into account, the present study tried to verify status of possession of lease *patta* land distributed by the government agencies in the study villages of both the districts.

Cross-verification was undertaken in the study villages in both the districts where lease *pattas* were granted to villagers. This cross-check covered name of the village, number of lease *patta* recorded, number of households verified, number of homestead or agricultural land, average area of land, year of possession, number of receipt of *patta,* number of physical possession and number of cases where location of land not shown as indicated in Table 7.2 and Table 7.3 which reflects the cross-checking position about lease of *patta* in Koraput and Rayagada districts respectively.

Table 7.2: Position of lease *patta* in Koraput district

Name of the village	No. of lease patta recorded	No. of HHs verified	No. of home-stead/ agric-ulture land	Average area of the land per head	Year of posse-ssion	No. of receipt of patta	No. of Phys-ical posse-ssion	No. of cases where places not shown
1	2	3	4	5	6	7	8	9
Devighat	38	35	H-31 A-02	0.02- 0.04	1986-87, 1998-99, 2004	35	04	27

Source: Tahasildar office, Koraput and field verification.

Regarding issue of *patta* for lease land, the position of Devighat village only was verified in Koraput district as a sample village. In Devighat village, 35 HHs were cross-checked. Out of this, 33 households received *patta* (homestead) measuring 0.02 to 0.04 acres and two HHs received agriculture land of 2 acres each. But as noticed in the field, except 4 HHs, no HH was shown land by demarcation/identification as a result of which, they are not in possession of actually allotted land. But in case of agricultural land, all are in possession of verified lands.

Table 7.3: Position of lease *patta* in Rayagada district

Name of the village	No. of lease patta recorded	No. of HHs verified	No. of home-stead/ agric-ulture land	Average area of the land per head	Year of posse-ssion	No. of receipt of patta	No. of Phys-ical posse-ssion	No. of cases where places not shown
1	2	3	4	5	6	7	8	9
Tadama	44	17	H-30 A-14	0.02ac. home-stead & 2ac. agrl.	2005-09	44	44	Nil

Name of the village	No. of lease patta recorded	No. of HHs verified	No. of home-stead/ agric-ulture land	Average area of the land per head	Year of posse-ssion	No. of receipt of patta	No. of Phys-ical posse-ssion	No. of cases where places not shown
1	2	3	4	5	6	7	8	9
Matikona	49	21	H-49 A-00	0.25ac. homestead	2009	49	49	Nil
Gunakhala	80	28	H-0 A-80	D-3 3 ac.	2010	80	80	Nil
Bariguda	16	11	H-0 A-16	D-3 2ac.	1998-2009	16	11	5
Gujalpadu	93	26	H-71 A-22	D1,D2 & D-3 2ac. & 0.02 hometead	1982	00	00	93
K.Dengasari	28 (Including T.Dengasari)	13	H-05 A-23	0.25 home-stead	2006	28	28	Nil
Bhatapur	160	60	H-67 A-93	D-3 2 ac.	2009-10	160	157	03
Majhihalma	26	06	H-26 A-0	0.01 homestead	1999	26	26	Nil
Garedi	12	12	H-0 A-12	0.27		00	00	12
Total	508	194	H-248 A-260			403	395	113

Source: Tahasildar office, Rayagada and field verification.

In Rayagada district, position of 9 study villages was verified for the purpose. From the Table 7.3, it transpires that 395 households physically possessed lands and 113 households even had not seen their lands for which *pattas* were granted.

Restoration of Illegally Transferred Land

As regards restoration of tribal land, only two cases were

found and checked in the study villages i.e., one in Devighat village and another in village Mania in Koraput district. In Rayagada district, details about lease land were checked in 9 study villages. Out of 508 total recorded households (248 homestead and 260 agricultural land during the period (1982-2010), 194 households were verified. Table 7.4 indicates the position of restoration of lands in Koraput district.

Table 7.4: Restoration of tribal land in study villages of Koraput district

Name of the village	Name of the opposition party	Name of the 1st party	Khata No.	Plot No.	Area in acres	Year of restoration	Whether any information about restoration	Whether physically possessed
1	2	3	4	5	6	7	8	9
Devighat	Manoranjan Senapaty S/o Radhanath Devighat	Dasarathi Jani S/o Jheruku Devighat	11	38	1.9	2005	No	No
Mania	Sankarlal Agrawal S/o Babulal, Semiliguda	Ghenu Lamta, S/o Mulia, Talamania	63/ 131	167	3.03	2007	No	No

Source: OSD section, Sub-collector office, Koraput and field verification.

As per the government record, the lands of two tribal persons only were restored, one in Devighat on 15.04.05 and another in Mania on 13.04.07 in Koraput district. But they reported that they have neither been informed nor are in possession of these lands. The opposite parties are still in possession of the said lands.

Table 7.5: Restoration of tribal land in study villages of Rayagada district

Name of the village	Name of the opposition party	Name of the 1st party	Khata No.	Plot No.	Area in acres	Year of restoration	Whether any information about restoration	Whether physically possessed
1	2	3	4	5	6	7	8	9
Matikona	Sahukar Bhima	Himirika Puniga	5/4	58/2	0.02	2002	No	No
	Krushna Sahukar	Himirika Reli	15	57	0.02	2002	No	No
	Nimalpuri Simanchal	Himirika Puniga	5/ka	52/2	0.02	2002	No	No
	Nimalpuri Ramarao	Himirika Bhima	8	48	0.02	2001	No	No
	Nimalpuri Sankarrao	Himirika Bhima	8	48	0.02	2001	No	No

Source: Sub-collector office, Rayagada and field verification.

The same is the case in Rayagada district. The original land owners whose lands were restored as per government record were not informed about such restoration (shown in Table 7.5). When the tribal victims were asked about their interest in getting their land back, they abruptly said that they have sold the lands to the parties who have already constructed houses on these lands and are occupying at present. Since they (transferees) have no other houses to stay, it would be an injustice for them to claim the transferred land for restoration back in their names.

Suggestions

The following suggestions emerged from the opinions given by the key stakeholders with regard to the effective implementation of the Act.

- Sri Gupteswar Panigrahy, an eminent lawyer working for more than 20 years on land issues suggested that the tribal person who is in need of money and wants to dispose off his land, has to intimate the same in writing to the competent authority. It would be for the government officer to go for auction for the land so that the tribal gets reasonable price for the land. Fraudulent practices can be reduced in this way. At the same time, the tribal person can get justice and meet his urgent needs.
- It was mentioned by Mr. Surendra Karan, an expert on land issues who worked on land rights for a longtime in Koraput District now having his own organization namely CDRT, Koraput that the OSATIP, 1956 (Regulation 2) as amended in 2002, has done injustice to the tribal people to some extent by restricting their freedom to dispose off immovable lands under compelling circumstances. Based on his experience in tribal areas, he suggested that a Village Land Committee for land should be constituted in each village. That committee will enquire whether the particular tribal person is in need of money for genuine reason, and then the matter will go to Gram Sabha. If Gram Sabha approves for the transfer, then only transaction can take place. Transaction price should also be decided in the Gram Sabha meeting. This method may reduce fraudulent transfer transactions.
- It was suggested by Jagannath Hembram, RI Nighamaniguda, that all illegal transfers of land cannot be detected by RIs. Sarpanches, ward members and PS members can only detect such cases in their respective areas, which could be intimated to RIs concerned for further action. Even NGOs in their respective operational areas can intervene and detect illegal transfers. In the process, people will have a genuine fear for resorting to illegal transactions.
- Marli Drulia, Sarpanch of Rajput suggested that (i) all land transactions should be intimated to the concerned Sarpanches, (ii) the proposed sale of land should be placed in Gram Sabha/Palli Sabha/GP meeting before

submitting the final sale deeds, (iii) for effective implementation of the 1956 Regulation (Regulation 2), awareness about the regulation should be generated among the people and the elected representatives as well.

- Dr. P.C. Mohapatra, COATS, Koraput said that as a result of application of 2002 amended regulation, transfer of immovable property by scheduled tribes to non-scheduled tribes has now been completely banned. However, restoration of tribal lands in respect of transactions where prior permission was not taken from appropriate authority under Regulation (2) of 1956, has to be strictly followed. In exceptional cases, where the landholder is really in need of money having no alternative with him, should inform the concerned authority and obtain permission for calling for sale through public auction of the land. Dr. Mohapatra expressed that there could be no confidence on the efficiency of Palli Sabha and Gram Sabha injudiciously according approval in such matters.
- Mr. Jhari Sethi, RI Panchada asserted about the proper working of the Regulation 2 and held that restoration of illegally transferred lands is being done perfectly. The original landowners are getting their lands back and cultivating them as before.
- Mr. Bhagat Jani, RI Kakiriguma suggested the following for better implementation of the regulation:
  (a) Instead of total ban on transaction of land from tribal to non-tribal, the Regulation should permit STs to transfer land to the non-tribals by retaining at least 10 acres at their disposal. In the process, STs can dispose of their land in excess of 10 acres to meet emergent situations.
  (b) Community level awareness programmes should be undertaken for effective implementation of the regulation.
- Smt. Geeta Muduli, Sarpanch Padmapur mentioned that Sarpanches should be informed of any land acquisition and their approval obtained prior to acquiring any land in the locality. She cited an instance

in her locality where one Bihari person acquired a small area without taking her consent. When she asked about the same, the person clarified that such permission was taken from the previous Sarpanch who is now dead. Smt. Muduli also informed that she had no idea about the intervention of Gram Sabha in land alienation and other land related issues where Gram Sabha has the power to give consent for transfer and restoration of such alienated lands. All these go to indicate that due to lack of awareness about the Acts/laws/regulations, the implications of the laws are not being understood by the people properly. Training and awareness programme is highly essential for effective implementation of the laws.

- Mr. Bidyut Mohanty of SPREAD, a local NGO working on tribal land issues since 1994 in Koraput district mentioned that in Koraput, about 40 per cent to 50 per cent lands are government lands. Recently, few homestead lands were settled and *pattas* have been issued. The victims of some of the displaced families yet continue to stay in forests. They have not been settled. Under Vasundhara Scheme, land has been allotted to tribals. But it is only in pen and paper. Physical possession of these lands has not been given to the allotees. As regards the 2002 amendment regulation of (OSATIP, 1956, Regulation 2), he opined that the regulation is good and effective. As a result of this amendment, illegal transaction of tribal lands has to a great extent been stopped. This new provision of law has very good objectives to strengthen livelihood sustainability of the tribals. There should not be any relaxation in the existing provision. Regarding the question of emergency needs like medical treatment etc., it is the responsibility of the government to help the poor tribals. Since land is the main source of livelihood, alienation of tribal lands, in any form, should be prevented as far as practicable.
- Mr. Subas Rait, Tahsildar Koraput was emphatically of the view that absence of survey and settlement is the main cause for all disputes relating to land. Since

RoRs have not been updated, it is very difficult on the part of government to settle mutation cases for which people are more interested. Since the number of living heirs has increased manifold and the process is lengthy, it is very difficult to settle all cases. But mutation for new purchases may have no problem. Regarding restoration he affirmatively said that the instituted cases are being disposed of leading to restoration. But after restoration, government functionaries are not in a position to verify physical possession of the original landowners. About relaxation required in the Act, he clarified that if lands are transferred from tribals, government do not have sufficient land for allotment to tribals. Hence this Act definitely is good for the benefit of the tribals. But the Act should be more effective, and more and more sensitization programmes should be conducted in the villages. Finally, he suggested that awareness programmes on survey and settlement can solve maximum problems relating to land.

- Mr. Biranchi Narayana Rath, Sub-Collector, Koraput emphasized on the relevance of 2002 amendment regulation. He strongly stated that the provision is stringent in all respects. He informed that total number of cases instituted in the district is 391 out of which 284 cases were registered on application and 107 cases on *suo motu* cognizance. Out of these, in 182 cases lands have been restored. He suggested that NGOs should take initiative as facilitators for smooth implementation of the provisions of the regulation. As all illegal cases are not being brought to their notice and both the parties do the transactions mutually for their own benefit, it is difficult to trace-out all cases. This can be avoided through NGOs' intervention. More and more trainings should be provided to the people for smooth implementation of the provisions of the amended regulation.
- Mr. Digvijayanath Padhi, Sub-Collector, Rayagada, pointed out that the cases which are registered under 3(B) 2 were to be applied in form-II before 2005. But

those who did file claims for their lands were restored on *suo mottu* action. This was the right direction as per his views. He strongly pleaded for enforcing the working of the provisions of 1956, Regulation (2). The 2002 amendment is perfectly all right and there is no need to change it, he said. This regulation is for the well-being of the tribals.

- The RI, Chanchadaguda of Rayagada district said that 2002 amendment is certainly good for the tribals. But in genuine cases, this should be relaxed particularly in case of those having more lands. Otherwise, all types of transaction should be banned.
- With regard to the gaps in 2002 Amendment Act read with OSATIP, 1956, Regulation (2), Advocate, Sri B.S. Nayak, Rayagada mentioned that if the land is to be utilized for public purpose, then the competent authority has to grant permission to transfer in favour of the concerned Institution. The Act as a whole is beneficial to the tribal people.
- RI Bhatpur of Rayagada district opined that the amended provisions are stringent in the interest of tribals and there is no need for any further change. The legislation is intended for the welfare and sustainable livelihood of the tribals.
- Sarapanch, Dumuriguda of Rayagada stated that Gram Sabha is not discharging any role in settling land related disputes.

Key Findings

- OSATIP, 1956, Regulation (2) followed by 2002 amendment further restricts transfer of tribal land. The objectives of these regulations are very good for ensuring better livelihood sustainability of the tribals. At the same time, the regulations deprive poor tribals to sale their land even in compelling/emergent situations when they face financial difficulties
- The Act is not effective in cases where transaction takes place without any written evidence, which is otherwise called verbal transfer.

- Lack of awareness about the regulation also creates problems in its effective implementation. Hence, awareness about different land related laws especially laws preventing transfer of tribal lands should be generated among the community.
- The role of Gram Sabha in land alienation under PESA is not effective.
- Sarpanches are not aware of PESA where role of Gram Sabha is very crucial.
- During cross-verification with reference to government records in Koraput district, it came to notice that in some cases of restoration of tribal lands, lands have been restored back but the land owners have not been informed about such restoration and therefore, are not in actual physical possession of the lands.
- Government have distributed *patta* in respect of lease lands but the majority of *patta* holders are not in physical possession of those lands as they have not been shown the land physically or physical possession of land has not been effected.
- In case of Rayagada district, the transferees are still in possession of those lands. When the tribal victims were asked about their interest in getting back their land, they abruptly said that they sold lands to the parties who have already constructed houses and staying on these lands at present. As the transferees have no other houses to stay, it would be an injustice to claim for return of the sold lands.
- The regulation should be made more effective through proper and regular monitoring during its implementation.
- Further amendment to the regulation is needed in the interest of the tribals to meet emergent situations.
- It is very difficult on the part of government to update the mutation cases because the number of legal heirs has increased even to more than 50 in a family in some cases. As a result, mutation cases cannot be settled even at government level. So survey and settlement operation should be done immediately for updating record of rights to remove confusion and uncertainty over tenurial rights.

Appendix-I

Orissa Scheduled Areas Transfer of Immovable Property (OSATIP) 1956, Regulation-2

[1] THE ORISSA SCHEDULED AREAS TRANSFER OF IMMOVABLE PROPERTY (BY SCHEDULED TRIBES) REGULATIONS, 1956

A. Regulation

[A Regulation to control and check transfer of immovable property in the Scheduled Areas of the State of Orissa by Scheduled Tribes]

Whereas, it is expedient to control and check transfers of immovable property by the Scheduled Tribes in the Scheduled Areas of the State of Orissa.

Now, therefore, in exercise of the powers conferred by Sub-paragraph (2) of Paragraph 5 of the Fifth Schedule to the Constitution, the Governor of Orissa, is pleased to promulgate following Regulations made by him in the Seventh Year of Republic of India:

1. Short title, extent and commencement

1. These Regulations may be called ***the Orissa Scheduled Areas Transfer of Immovable Property (By Scheduled Tribes) Regulations, 1956.***
2. It extends to all the Scheduled Areas of the State of Orissa.
3. It shall come into force at once.

2. Definitions

In this Regulation unless there is anything repugnant in the subject or context –

[2](a) *"agriculture"* includes making land fit for cultivation. cultivation of land, improvement of land including development of sources of irrigation, raising and harvesting of crops, horticulture, forestry, planting of trees, farming, cattle breading, dairy farming, seed farming, pisciculture, agriculture, sericulture, piggery, poultry farming and such other activities as are generally carried on by agriculturists, dairy farmers, cattle breeders, poultry farmers and other category of persons engaged in similar activities including marketing of agricultural products, their storage and transport and the acquisition of implements and machinery in connection with any such activity, and agricultural purpose shall be construed accordingly, (x x x)[3]

[4][(a-b) *"competent authority"* means the Collector and any other Officer or Officers appointed by the State Government by notification to perform all or any of functions of a competent authority under this regulation]

[5][(b) "*Grama Panchayat*" and "*Grama Sasan*" respectively mean the Grama Panchayat and Grama Sasan as defined in the Orissa Grama Panchayat Act, 1964;]

(c) *"Immovable property"* does not include standing timber growing crops or grass,

[6][(d-l) *"public financial institution"* means,–

(i) Any bank to which the Orissa Co-operative Societies Act, 1962 (Orissa Act 2 of 1963) applies.

(ii) A banking company within the meaning of Banking Regulation Act 10 of 1989.

(iii) The State Bank of India constituted under the Bank of India Act 23 of 1955.

(iv) A subsidiary Bank within the meaning of the State Bank of India (Subsidary Bank) Act 33 of 1959;

(v) A corresponding new Bank within the meaning of

Banking Companies (Acquisition and Transfer Undertakings) Act, 5 of 1970 ;

(vi) The Agricultural Refinance Corporation established under the Agricultural Refinance Corporation Act, 1963 ;

(vii) The Agricultural Finance Corporation Limited, a company incorporated under the Companies Act 1 of 1956;

(viii) The Agro-Industries Corporation ;

(ix) Any other financial institutions which may be notified by the State Government;]

[7][(e) ***"Scheduled Areas"*** and ***"Scheduled Tribe"*** shall respectively means and Scheduled Areas specified in respect of the State of Orissa in the Scheduled Areas (State of Bihar, Gujarat, Madhya Pradesh and Orissa) Order, 1977 and the "Scheduled Tribes" specified in respect of the State of Orissa in the Constitution (Scheduled Tribes) Order, 1950 as modified from time to time.]

(e-1) [x x x][8]

Explanation: For the purposes of conversion, one acre shall be equal to 0.4047 hectare.

(f) ***"Transfer of immovable property"*** means mortgage with or without possession, lease sale, gift, exchange or any other dealings with such property not being a testamentary disposition and includes a charge or contract relating to such property.

3. Transfer of Immovable property by a member of the Scheduled Tribe

[9]["(1) Notwithstanding anything contained in any law for the time being in force any transfer of immovable property by a member of a Scheduled Tribe, except by way of mortgage executed in favour of any public financial institution for securing a loan granted by such *institution* for any agricultural purpose, shall be absolutely null and void and of no force or effect whatsoever, unless such transfer is made *in favour of* another member of a Scheduled Tribe :

Provided that –

(i) nothing in this sub-section shall be construed as to permit any member of a Scheduled Tribe or his successor-in-interest to transfer any immovable property which was settled with such member of Scheduled Tribe by or under any authority of the State or the Central Government or under any law for the time being in force;

(ii) in execution of any decree for realisation of the mortgage money, no property mortgaged as aforesaid shall be sold in favour of any person not being a member of a Scheduled Tribe; and

(iii) a member of a Scheduled Tribe shall not transfer any land if the total extent of his land remaining after the transfer will be reduced to less than two acres in case of irrigated land or five acres in case of unirrigated land.

Explanation I: For the purposes of this sub-section, a transfer of immovable property :

(a) in favour of a female member of a Scheduled Tribe, who is married to a person not belonging to any Scheduled Tribe, shall be deemed to be a transfer made in favour of a person not belonging to a Scheduled Tribe; and

(b) shall include a transfer of immovable property to a person belonging to a Scheduled Tribe for consideration paid or provided by another person not belonging to any such Tribe.

Explanation II: For the purposes of clause (iii) of the proviso, the expression "irrigated land" shall mean such land which is irrigated at least for one crop in a year, and the expression "unirrigated land" shall be construed accordingly; and

(2) Where a transfer of immovable property is made in contravention of Sub-sec. (i) the competent authority may, either on application by any one interested therein [10](or on information received from the Gram

panchayat) or on his own motion and after giving the parties, an opportunity of being heard order ejectment against any person in possession of the property claiming under the transfer and shall cause restoration of possession of such property or the transferor or his heirs. In causing such restoration of possession the competent authority may take such steps as may be necessary for securing compliance with the said order or preventing any breach of peace.

[11][**Provided that** if the competent authority is of the opinion that the restoration of possession of immovable property to the transfer his heirs is not reasonably practicable, he shall record his reasons thereof and shall, subject to the control of the State Government settle the said property with another member of Scheduled Tribe or in the absence of any such member, with any other person in accordance with the provisions contained in the Orissa Government Land Settlement Act, 33 of 1962.

[12][**"Provided further that** where it is decided by the competent authority to settle the said property with any person other than a person belonging to a Scheduled Tribe in accordance with the Orissa Government Land Settlement Act, 1962, he shall obtain prior approval of the concerned Grama Panchayat, accorded with the concurrence of the Grama Sasan.")

Explanation: Restoration of possession means actual delivery of possession by the competent authority to the transferer or his heirs.

(3) Subject to such conditions as may be prescribed an appeal if preferred within thirty days of the order under Sub-sec (2) shall if made by the Collector lie to the [13](Revenue Divisional Commissioner) and if made by any other competent authority to the Collector, or any other officer specially empowered by the State Government in this behalf.

[14][(3-a) *(i)* Any order passed by the competent authority under sub-section (1) or under sub-section (2) may, within a

period of five years from the date of such order, be revised, either on own motion or otherwise, by the Revenue Divisional Commissioner if the order was passed by the Collector and if it was passed by any other competent authority, after giving the parties concerned a reasonable opportunity of being heard :

Provided that no order shall be revised under this sub-section if an appeal against such order has been entertained under sub-section (3).

(a) For the purpose of revising any order, the Revenue Divisional Commissioner or, as the case may be, the Collector shall follow such procedure as may be prescribed and shall have power to call for and examine the records of the proceedings wherein such order was passed and pass such order as he deems fit.] ;

(4) Subject to the provisions of [15][any order passed under Sub-sec. (3) or Sub-sec. (3-a)] the decision of the competent authority under Sub-section (2) shall be final and shall not be challenged in Court of law.

[16][3-A. Eviction of Persons in Unauthorised Occupation Property

(1) Where a person is found to be in unauthorised occupation of any immovable property of a member of the Scheduled Tribes by way of trespass or otherwise the competent authority may, either on application by the owner or any person interested therein, (or on information received from the Grama Panchayat) or on his own motion, and after giving the parties, concerned an opportunity of being heard order ejectment of the person so found to be in unauthorised occupation and shall cause restoration of possession of such property to the said member of the Scheduled Tribes or to his heirs.

(2) The provisions contained in Sub-sections (2), (3) and (4) of Section 3 shall *mutatis mutandis* apply to the proceedings instituted or initiated under Sub-sec. (1).

[17]["(3) In every case after finalisation of the proceedings

under sub-section (1), the competent authority shall make a report to the concerned Grama Panchayat about the order of ejection passed in respect of any person in unauthorised occupation of any immovable property of a member of a Scheduled Tribe and the restoration of possession of the property to such member or his heirs and in case of failure of such restoration, the reasons for such failure.]"

[18][**"3-B. Reservation of land of members of Scheduled Tribes which was transferred by fraud —**

(1) Every person who, on the date of commencement of the Orissa Scheduled Areas Transfer of Immovable Property (By Scheduled Tribes) Amendment Regulation, 2000 (hereinafter referred to in this section as the Amendment Regulation of 2000), is in possession of agricultural land which belonged to a member of a Scheduled Tribe at any time during the period commencing on the 4th October, 1956 and ending on the date of commencement of the Amendment Regulation of 2000 shall, within two years of such commencement, notify to the Sub-Collector in such form and in such manner as may be prescribed, all the information as to how he has come in possession of such land.

(2) If any person fails to notify the information as required by sub-section (1) within the period specified therein it shall be presumed that such person has been in possession of the agricultural land without any lawful authority and the agricultural land shall on the expiration of the period aforesaid, revert to the person to whom it originally belonged and if that person be dead, to his heirs.

(3) On receipt of the information under sub-section (1), the Sub-Collector shall make such enquiry as may be necessary about all such transactions or transfer and if he finds that the member of Scheduled Tribe has been defrauded of his legitimate right he shall declare the transaction null and void and, —

(a) where no building or structure has been erected on the agricultural land prior to such finding, pass an order revesting the agricultural land in the transferor and if he be dead, in his heirs;

(b) where any building or structure has been erected on the agricultural land prior to such finding, he shall fix the price of such land in accordance with the principles laid down for fixation of price of land in the Land Acquisition Act, 1894 and order the person referred to in sub-section (1) to pay to the transferor the difference, if any, between the price so fixed and the price actually paid to the transferor.

Provided that where the building or structure has been erected after the commencement of the Amendment Regulation of 2000, the provisions of clause (b) shall not apply :

Provided further that fixation of price under clause (b) shall be with reference to the price on the date of registration of the case before the Sub-Collector.]

(4) Inadmissibility of a deed for the purpose of registration

Notwithstanding anything contained in the Indian Registration Act, XVI of 1908 no deed of transfer of any immovable property executed in contravention of provision of this Regulation shall be accepted for registration.

(5) Surrender or relinquishment not valid unless settled by landlord—

(1) No surrender or relinquishment of any holding or a part of a holding by a tenant to a landlord under any law for the time being in force applicable to such tenancy; such tenant being a member of a Scheduled Tribe, shall be valid unless after such surrender of relinquishment the landlord thereof by whatever name called either settles the said holding or part of the holding, as the case may be, with another member of a Scheduled Tribe or else retains it in his possession or settles it with any other person with the

approval of the competent authority when such member of a Scheduled Tribe is not available.

(2) Any surrender or relinquishment shall be deemed to be a transfer of immovable property within the meaning of this Regulation and except as otherwise provided in the Sub-section (1) the other provisions of this Regulation shall so far as may be applied.

(3) Nothing in this section shall apply to any tenant holding immediately under the State Government.

6. Bar to attachment of immovable property

In execution of a money decree against a member or a Scheduled Tribe, no right, title or interest held by him in any immovable property within any Scheduled Area shall be liable to be attached and sold except as and if prescribed.

[19][**7. Punishment for offences—**

(1) If any person is found to be in possession of any immovable property in contravention of the provisions of this Regulation, then, without prejudice to his liability to ejectment under this Regulation, or where any person, having been evicted under any provision of this Regulation from any immovable property belonging to a member of a Scheduled Tribe, continues to be in possession of the same, he shall be punishable with rigorous imprisonment for a term which may extend to two years, or with fine which may extend to five thousand rupees, or with both.

(2) Where any person, having been evicted under any provision of this Regulation from any immovable property belonging to a member of a Scheduled Tribe, reoccupies the same shall be punishable with rigorous imprisonment for a term, which may extend to three years, or with fine, which may extend to ten thousand rupees, or with both.

(3) When a Court imposes sentence of fine or a sentence of which fine forms a part, the Court may;

when passing judgment, order the whole or any part of the fine to be paid to the member of Scheduled Tribe to whom the immovable property belongs, or to his heirs.]

7-A. Offences to be tried by Executive Magistrate and to be cognizable—

(1) The State Government may confer, on a Executive, Magistrate, the powers of a Judicial Magistrate of the first class or of the second class for the trial of the offences under this Regulation and on such conferment of powers, the Executive Magistrate, on whom the powers are so conferred, shall be deemed, for the purposes of the Code of Criminal Procedure, 1973, to be a Judicial Magistrate of the first class or of the second class, as the case may be.

(2) An offence under this Regulation may be tried summarily by a Magistrate.

(3) Every offence under this Regulation shall be cognizable.

7.B. Consideration money not refundable in case of invalid transfer

Notwithstanding anything contained in any law for the time being in force, where transfer of immovable property is found to have been made in contravention of the provisions of Section 3 and the transferee or any other person in possession of the property has been evicted therefrom under the said section, the transferee shall not be entitled to the refund of any amount paid by him to the transferer by way of consideration for the transfer.

7.C. Burden of proof and power of Court to ignore admission—

Notwithstanding anything contained in any other law for the time being in force.

(a) If in any proceedings under this Regulation the validity of the transfer or relinquishment of any immovable property is called in question or if such proceedings are for the recovery of possession of immovable property, the burden of proving that the transfer of relinquishment was valid shall lie on the transferee [20][or any person claiming the right to the property through the transferee] :

(b) The Court shall in any suit or proceeding relating to the transfer of immovable property a member of the Scheduled Tribes have power to require in fact expressly or impliedly admitted by such member to be proved otherwise than by admission.

[21][7-D **Amendment of the Limitation Act, 1963 in its Application to the Scheduled Areas**

In the Limitation Act, 1968 in its application to the Scheduled Areas in the Schedule, after the words "twelve years" occurring in the second column against Art. 65, the words "twelve years" and figure "but thirty years" in relation to immovable property belonging to a member, of a Scheduled Tribe specified in respect of the State of Orissa in [22][the Constitution (Scheduled Tribes] Order, 1950 as modified from time to time shall be added.]

[23][7-E. **Bar of jurisdiction of Civil Courts**

No Civil Court shall have jurisdiction to try and decide any suit or proceeding so far as it relates to any manner which any officer or other competent authority is empowered by or under this Regulation to decide.]

8. **Rules –**

(1) The State Government may from time to time make Rules consistent, with the provisions of this Regulation to carry out the purposes thereof.

(2) All Rules made under this section shall be published in the Gazette and on such application shall have the effect as if enacted under this Regulation.

9. **Repeal –**

(1) On and from the date of commencement of this regulation shall stand repealed namely –

(a) The Agency Tracts Interests and Land Transfer Act, 1917 (Madras Act of 1917) :

(b) The enactments mentioned in Column 2 of the (Scheduled to the extent specified in

Column 3 thereof insofar as they are in force (in the Scheduled Areas).

(2) Notwithstanding such repeal by this regulation the repeal shall not effect–

(a) the previous operation of such law of provisions so repealed or validity affect or consequence of anything done or suffered thereunder ;

(b) any right, title, privilege, obligation or liability acquired, accrued or incurred under the said law or provision; or

(c) any investigation, legal proceedings or remedy in respect of such right, title, privilege obligation or liability; or

(d) any release or discharge from any debt, penalty, obligation, liability, claim or demand and any such investigation, legal proceeding or remedy may be instituted, continued or enforced and any such penalty, forfeiture of punishment may be imposed as if this Regulation had not been made.

Note: Madras Act 1 of 1917, Orissa Act IV of 1950 & Central Provinces Act XI of 1898 repealed by Regulation 1 of 1966.

REFERENCES

1. Received the assent of the President on the 21st September, 1956 and was first published in the Orissa Gazettee, Extraordinary, dated 4th October, 1956.
2. Added by Orissa Regulation 1 of 1975.
3. Sub-clause (aa) is Omitted vide Notifn. No. 11714-Legis. dt. 4.9.2002 by Regulation 1 of 2002 titled as OSATIP (by ST) Amendment Regulation 2000.
4. Re-numbered by Orissa Regulation 1 of 1975 and renumbered by Notification No. 7894, dt. 27.7.97. Regulation 1 of 1997.
5. Substd. by Regulation 1 of 2002 dt. 4.9.2002.
6. Added by Orissa Regulation 1 of 1975.
7. Substituted by Orissa Regulation 1 of 1979.
8. Omitted by Regulation 1 of 2002 dt. 4.9.2002.

9. Substituted by Orissa Regulation 1 of 2002 dt. 4.9.2002.
10. Added by Orissa Regulation 1 of 2001 dt. 4.9.2002.
11. Added by Orissa Regulation 1 of 1966.
12. Added by Orissa Regulation 1 of 2001 dt. 4.9.2002.
13. Substituted by Notification No. 7894 dt. 27.7.97.
14. Inserted by *ibid.*
15. Substituted by Notification No. 7894. dt. 27.7.97.
16. Added by Orissa Regulation 1 of 1975.
17. Added by Orissa Regulation 1 of 2002 dt. 4.9.2002.
18. Inserted by *ibid.*
19. Substituted by Orissa Regulation 1 of 2002 dt. 4.9.2002.
20. *Ibid.*
21. Inserted by Notification No. 7894 dt. 27.7.1997.
22. Which was inserted therein by Orissa regulation of 1975 shall be deemed to have been so inserted with effect from the 2nd day of October, 1973 (Orissa Regulation 1 of 1976).
23. Inserted by Notification No. 7894 dt. 27.7.1997, Regulation 1 of 1997.

Appendix-II

Panchayats (Extension to the Scheduled Areas) (PESA) Act, 1996

THE PROVISIONS OF THE PANCHAYATS (EXTENSION TO THE SCHEDULED AREAS) ACT, 1996 NO. 40 OF 1996 (24TH DECEMBER, 1996)

An Act to provide for the extension of the provisions of Part IX of the Constitution relating to the Panchayats to the Scheduled Areas.

Be it enacted by Parliament in the Forty-seventh Year of the Republic of India as follows :—

Short Title

1. This Act may be called the Provisions of the Panchayats (Extension to the Scheduled Areas) Act, 1996.

Definition

2. In this Act, unless the context otherwise requires, "Scheduled Areas" means the Scheduled Areas as referred to in Clause (1) of Article 244 of the Constitution.

Extension of Part IX of The Constitution

3. The provision of Part IX of the Constitution relating to Panchayats are hereby extended to the Scheduled Areas subject to such exceptions and modifications as are provided in section 4.

Exceptions and Modifications to Part IX of the Constitution

4. Notwithstanding anything contained under Part IX of the Constitution, the Legislature of a State shall not make any law under that Part which is inconsistent with any of the following features, namely:—
   (a) a State legislation on the Panchayats that may be made shall be in consonance with the customary law, social and religious practices and traditional management practices of community resources;
   (b) a village shall ordinarily consist of a habitation or a group of habitations or a hamlet or a group of hamlets comprising a community and managing its affairs in accordance with traditions and customs;
   (c) every village shall have a Gram Sabha consisting of persons whose names are included in the electoral rolls for the Panchayat at the village level;
   (d) every Gram Sabha shall be competent to safeguard and preserve the traditions and customs of the people, their cultural identity, community resources and the customary mode of dispute resolution;
   (e) every Gram Sabha shall—
      (i) approve of the plans, programmes and projects for social and economic development before such plans, programmes and projects are taken up for implementation by the Panchayat at the village level;
      (ii) be responsible for the identification or selection of persons as beneficiaries under the poverty alleviation and other programmes;
   (f) every Panchayat at the village level shall be required to obtain from the Gram Sabha a certification of utilisation of funds by that Panchayat for the plans, programmes and projects referred to in clause (e);
   (g) the reservation of seats in the Scheduled Areas at every Panchayat shall be in proportion to the

population of the communities in that Panchayat for whom reservation is sought to be given under Part IX of the Constitution:

Provided that the reservation for the Scheduled Tribes shall not be less than one-half of the total number of seats: Provided further that all seats of Chairpersons of Panchayats at all levels shall be reserved for the Scheduled Tribes;

(h) the State Government may nominate persons belonging to such Scheduled Tribes as have no representation in the Panchayat at the intermediate level or the Panchayat at the district level:

Provided that such nomination shall not exceed one-tenth of the total members to be elected in that Panchayat;

(i) the Gram Sabha or the Panchayats at the appropriate level shall be consulted before making the acquisition of land in the Scheduled Areas for development projects and before re-setling or rehabilitating persons affected by such projects in the Scheduled Areas; the actual planning and implementation of the projects in the Scheduled Areas shall be coordinated at the State level;

(j) planning and management of minor water bodies in the Scheduled Areas shall be entrusted to Panchayats at the appropriate level;

(k) the recommendations of the Gram Sabha or the Panchayats at the appropriate level shall be made mandatory prior to grant of prospecting licence or mining lease for minor minerals in the Scheduled Areas;

(l) the prior recommendation of the Gram Sabha or the Panchayats at the appropriate level shall be made mandatory for grant of concession for the exploitation of minor minerals by auction;

(m) while endowing Panchayats in the Scheduled Areas with such powers and authority as may be necessary to enable them to function as institutions of self-government, a State Legislature shall ensure that the

Panchayats at the appropriate level and the Gram Sabha are endowed specifically with—

(i) the power to enforce prohibition or to regulate or restrict the sale and consumption of any intoxicant;

(ii) the ownership of minor forest produce;

(iii) the power to prevent alienation of land in the Scheduled Areas and to take appropriate action to restore any unlawfully alienated land of a Scheduled Tribe;

(iv) the power to manage village markets by whatever name called;

(v) the power to exercise control over money-lending to the Scheduled Tribes;

(vi) the power to exercise control over institutions and functionaries in all social sectors;

(vii) the power to control over local plans and resources for such plans including tribal sub-plans;

(n) the State Legislations that may endow Panchayats with powers and authority as may be necessary to enable them to function as institutions of self-government shall contain safeguards to ensure that Panchayats at the higher level do not assume the powers and authority of any Panchayat at the lower level or of the Gram Sabha;

(o) the State Legislature shall endeavour to follow the pattern of the Sixth Schedule to the Constitution while designing the administrative arrangements in the Panchayats at district levels in the Scheduled Areas.

Continuance of Existing Laws on Panchayats

5. Notwithstanding anything in Part IX of the Constitution with exceptions and modifications made by this Act, any provision of any law relating to Panchayats in force in the Scheduled Areas, immediately before the date on which this Act receives the assent of the President, which is inconsistent with

the provisions of Part IX with such exceptions and modifications shall continue to be in force until amended or repealed by a competgent Legislature or other competent authority or until the expiration of one year from the date on which this Act receives the assent of the President:

Provided that all the Panchayats existing immediately before such date shall continue till the expiration of their duration unless sooner dissolved by a resolution passed to that effect by the Legislative Assembly of that State or, in the case of a State having Legislative Council, by each House of the Legislature of that State.

Appendix-III

Orissa Resettlement and Rehabilitation Policy, 2006

In order to ensure sustained development through a participatory and transparent process, Government has framed a comprehensive resettlement and rehabilitation policy. Basic objectives of the policy are:

1. to avoid displacement wherever possible and minimize it exercising available options otherwise,
2. to recognize voices of displaced communities emphasizing the needs of the indigenous communities and vulnerable sections,
3. to ensure environmental sustainability through participatory and transparent process, and
4. to help guiding the process of developing institutional mechanism for implementation, monitoring, conflict resolution and grievance redressal.

With the above objectives in view, government has been pleased to lay down the following policy guidelines :

ORISSA RESETTLEMENT AND REHABILITATION POLICY, 2006

Preamble

Government of Orissa has been pursuing various development initiatives to improve the quality of lives. Ensuring social justice being one of the major cornerstones of development, the Government always proactively tries to make sure people's participation in development process. In spite of Government's intention to bring development to the people, development interventions do at times create undesirable consequences. Displacement due to large development projects is one such phenomenon. Government of Orissa has been responding to this problem through various projects specific Resettlement and Rehabilitation Policies and Plans. The current intervention of policy formulation has actually taken note of the lessons learnt through these past policies, which essentially reflects Government's genuine spirit of learning and retrospection. The present policy draws its strength from experiences from the implementation of past policies, best practices in other States and Orissa Government's Industrial Policy Resolution, 2001. Consultation with various direct and indirect stakeholders including civil society of the State has been conducted, and the views of the academicians and specialists in the field cf resettlement and rehabilitation have been considered as a part of democratic response of the Government in Policy formulation. Limitations of the past policies have been acknowledged and analyzed and a flexible framework has been attempted, which nonetheless demonstrates the dynamism of the Government. Unlike many other policies, there is a strong focus on the modalities of implementation of this Policy that makes it a vibrant instrument to promote sustainable development in the State.

2. Short Title, Application and Commencement

(i) This Policy may be called as "The Orissa Resettlement and Rehabilitation Policy, 2006" and shall come into effect from the date of its publication in the *Orissa Gazette.*

(ii) It shall apply to all those projects, for which acquisition of private land under Land Acquisition Act, 1894 or under any other laws for the time being in force or proclamation inviting objections in case of Government land is notified.

(iii) This shall also be applicable to all projects for which land is acquired through negotiation under the provisions of this Policy.

3. Definitions

In this Policy unless the context requires otherwise:

(a) **"Agricultural Land"** means land used or capable of being used for raising of crops, grass or garden produce, horticulture, dairy farming, fish farming, breeding and keeping of livestock and used as pasture or for any other purpose where such use is ancillary to agriculture.

(b) **"Compensation"** has the same meaning as assigned to it under the Land Acquisition Act, 1894.

(c) **"Cut-off Date"** for the purpose of compensation shall be the date on which the notification declaring the intention to acquire land under the relevant Act or under the provisions of this Policy is published.

Note: For the purpose of declaring eligibility for R & R benefits, the list of displaced families, will be updated on the 1st of January of the year in which physical displacement is scheduled to take place provided that those families who move into the project area after determination of the "cut-off- date", will not be eligible for any benefit.

(d) **"Displaced Family"** means a family ordinarily residing in the project area prior to the date of publication of notification under the provisions of the relevant Act and on account of acquisition of his/her homestead land is displaced from such area or required to be displaced.

(e) **"District Compensation Advisory Committee (DCAC)"** means the Committee constituted by Government under relevant provisions of this Policy.

(f) **"Family"** means the person and his or her spouse, minor sons, unmarried daughters, minor brothers or unmarried sisters, father, mother and other members residing with him or her and dependent on him or her for his/her livelihoods.

Note: Each of the following categories will be treated as a separate family for the purpose of extending rehabilitation benefits under this Policy.

(i) A major son irrespective of his marital status.

(ii) Unmarried daughtersister more than 30 years of age.

(iii) Physically and mentally challenged person irrespective of age and sex; (duly certified by the authorized Medical Board). For this purpose, the blind/the deaf/the orthopaedically handicapped/mentally challenged person suffering from more than 40% permanent disability will only be considered as separate family.

(iv) Minor orphan, who has lost both his/her parents.

(v) A widow or a woman divorcee.

(g) **"Government"** means the Government of Orissa in Revenue Department.

(h) **"Land Acquisition Officer (LAO)"** means an Officer appointed by the Government by an order to perform duties as such under Land Acquisition Act, 1894, for the project and shall also include a Special Land Acquisition Officer.

(i) **"Non-Government Organization (NGO)"** means any organization duly registered under the Society Registration Act, 1860 and functioning for public cause outside the Government.

(j) **"Original Family"** means the family, which at the time of Notification under provisions of relevant Act(s) is living together in a single household with a common kitchen.

(k) **"Periphery"** means the district(s) in which the project is geographically situated.

(l) **"Project"** means the construction, extension or improvement of any work such as reservoir, dam, canal, highway, industrial plant, factory, mining, national park, sanctuary, etc. as notified by the Government from time to time and includes its offices and establishment within the State.

(m) **"Project area"** for the purpose of extending R & R benefits means the land, which is acquired/alienated/purchased for establishment of any project.

(n) **"Project Director Resettlement & Rehabilitation (PD-RR)"** means an Officer appointed by the

Government by an order to perform such duties under R & R Policy of the State, for the project.

(o) **"Rehabilitation & Periphery Development Advisory Committee (RPDAC)"** means the Committee constituted by the Government under relevant provisions of this Policy by Government to look after rehabilitation and periphery development matters.

(p) **"State Level Compensation Advisory Committee"** means a Committee constituted by Government under the relevant provisions of this Policy.

(q) **"State Level Council on Resettlement & Rehabilitation (SLCRR)"** means the Council headed by the Chief Minister constituted by a notification of Government to that effect.

4. Policy Objectives

Objective of the Policy of the Government in general shall be:

(a) To avoid displacement where possible and minimize it, exercising available options otherwise.

(b) To facilitate resettlement/rehabilitation process:

   (i) Recognizing voices of the displaced communities (emphasizing the needs of the indigenous communities and vulnerable sections); and

   (ii) Ensuring environmental sustainability through participatory and transparent process; and

(c) To help guiding the process of developing institutional mechanisms for implementation, monitoring, conflict resolution and grievance redressal.

5. Survey and Identification of Displaced Families

(a) Ordinarily within two months of publication of notice for acquisition of land for the development project, a socio-economic survey would be undertaken in the manner to be decided by the Government for identification of displaced families and for preparing their socio-economic baseline.

(b) The list of displaced families shall be placed before and approved by the respective RPDAC.

(c) The list of displaced families so approved will be displayed at Collectorate/Block/Tehsil/ Panchayat and other conspicuous locations for wider dissemination.

(d) RDC shall realistically assess the requirement of land for acquisition before issue of notification under the relevant law(s) or under the provisions of this Policy.

(e) A socio-cultural, resource mapping and infrastructural survey shall be conducted by an independent agency to be identified by the Government to ensure proper benchmarking. It will be the responsibility of the concerned Project Authority to conduct the survey within two months of notifying the intention to acquire land under the provisions of the relevant law(s) or under the provisions of this Policy.

(f) A comprehensive communication plan for awareness creation shall be formulated and executed in the affected area. The detailed modalities of this exercise that include involvement of civil society will be notified by the Government. The cost of implementation of this communication plan shall be borne by the project(s).

(g) Gram Sabha or Panchayats at the appropriate level shall be consulted in scheduled areas before initiating Land Acquisition Proposal.

(h) Normal development programmes implemented by different agencies should be dovetailed with resettlement and rehabilitation package in resettlement habitats and made available to the displaced community on a priority basis.

(i) An identity card shall be issued to each displaced family in a manner prescribed by Government.

6. Project Types

For the purpose of R & R benefits under this Policy, Development Projects are classified into the following types:

A. Industrial Projects;
B. Mining Projects;

C. Irrigation Projects, National Parks and Sanctuaries;
D. Urban Projects and Linear Projects like roads and railways, power lines; and
E. Any other Projects.

7. Land Acquisition and Payment of Compensation/Award

Procedure prescribed by Government shall be followed in acquiring land and other property and for payment of compensation/award. All compensation money due to the "displaced families" shall be paid through account payee cheques. As regards "public property" like School Building, Club House, Hospital, Panchayat Ghar, electrical installation, place(s) of worship, value of such property affected shall be deposited with the concerned District Collector. Either Project or District Administration shall take up construction at the place as would be determined in consultation with representatives of displaced persons.

The Project proponent may opt for direct purchase of land on the basis of negotiated price after issue of notification requiring acquisition of land under relevant Act(s). If acquisition of land through direct purchase fails, other provisions of the relevant Act may be invoked.

Land not utilized by the Project within the prescribed time limit and for the required purposes shall be resumed.

8. Resettlement and Rehabilitation Plan

Based on the list approved by Government and option of displaced families, Resettlement and Rehabilitation Plan shall be prepared by the Collector for resettlement and rehabilitation after due consultation with displaced communities in the manner determined by the Government. Such plan should address the specific needs of the women, vulnerable groups and indigenous communities. The same will be placed before the RPDAC for approval.

While preparing the plan, the following aspects should be taken into consideration:

(i) Site for the resettlement habitat shall be selected by the RPDAC in consultation with the displaced families.

(ii) No physical displacement shall be made before the completion of resettlement work as approved by the RPDAC. The certificate of completion of resettlement work will be issued by the Collector.

(iii) Gram Sabha shall be consulted.

(iv) Where there is multiple displacement additional compensation amounting to 50% of the normal compensation payable, shall be paid to each displaced family over and above the normal compensation in form of *ex gratia*.

(v) Provisions relating to rehabilitation will be given effect from the date of actual vacation of the land.

(vi) Project Authority shall abide by the provisions laid down in this Policy and the decisions taken by RPDAC from time to time provided they are within the ambits of the approved Policy of the Government.

(vii) District Administration and Project Authorities shall be jointly responsible for ensuring that the benefits of R & R reach the target beneficiaries in a time bound manner.

(viii) Record of Rights of the land and houses allotted to the displaced persons should be handed over to them by the District Administration while resettling them in the Resettlement habitat. The District Administration shall take steps for immediate declaration of the new Resettlement habitat as a Revenue Village if it is not a part of an already existing Revenue Village.

(ix) Steps will be taken by the Project Authorities for acclimatization of the resettled people in new habitat including development of cordial social relationship between the host and resettled communities and to ensure as far as practicable overall improvement of standard of living of the displaced families.

(x) Subject to the details regarding provision of employment as enunciated elsewhere in the Policy the project authorities shall give preference in the matter of employment, both direct and indirect as well as through contractors employed by them, for execution, operation and maintenance of the project, to local persons as per the detailed guidelines issued by the State Government from time to time.

9. Rehabilitation Assistance

Rehabilitation Assistance will be specific to the 'type' of project as mentioned at Para. 5 above, because of difference in nature of projects, their source(s) of funding and magnitude of displacement/impact.

I. Type A: Industrial Projects

(a) **Employment:** Families shall be eligible for employment, by the project causing displacement or loss of all agricultural land. For the purpose of employment, each original family will nominate one member of such family. However, the families as mentioned at Para. 2 (f), (i), (ii), (iii), (iv) or (v) will not be considered separately for employment. Any one from among these categories may, subject to eligibility, be nominated by the family as defined in Para. 2 (f) for the purpose of employment. The project proponent will give preference to the nominated members of the displaced/other families in the matter of employment. The order of preference will be as follows:

(i) Displaced families losing all land including homestead land,

(ii) Displaced families losing more than 2/3rd of agricultural land and homestead land,

(iii) Displaced families losing more than 1/3rd of agricultural land and homestead land,

(iv) Displaced families losing only homestead land but not agricultural land,

(v) Families losing all agricultural land but not homestead land.

The Project Authority will make special efforts to facilitate skill up-gradation of the nominated member of each displaced/other family to make him/her 'employable' in their project.

1. In case of nominees of displaced/other families eligible for employment otherwise, the upper age limit shall be relaxed by five years.
2. Project Authorities should notify their employment capacity sufficiently in advance.

3. As far as practicable, the objective shall be to provide one member from each displaced/other family as mentioned above with employment in the project. However, where the same cannot be provided because of reason to be explained in writing, cash compensation as mentioned below shall be provided to the displaced/ other families. Displacedother families, who do not opt for employment/self-employment as mentioned in sub-Para. (a) and (b), shall be provided by the Project Authority with one time cash assistance in lieu of employment at the scale indicated below:

Categories of families referred to in clauses (i) to (v) of Para. 2 (f) shall not be considered separately for employment in the project. However, they will be paid one time cash compensation in lieu of employment:

Provided that if any industry is willing to offer employment to more than one member of a family, it may do so in lieu of one-time cash assistance:

Displaced/other families who are eligible to be considered for employment and families defined under claues (i) to (v) shall be provided with one time cash assistance in lieu of employment/self-employment in the scale indicated below:

Sl. No.	Families under category as per sub-Para. (a) above	Amount of one time cash assistance (Rs. in lakhs)
(i)	Displaced Families coming under category (i)	5.00
(ii)	Displaced Families coming under category (ii)	3.00
(iii)	Displaced Families coming under category (iii)	2.00
(iv)	Displaced Families coming under category (iv) and (v)	1.00

(b) **Training for Self-employment:** Project Authority under the guidance of the Collector concerned will make adequate arrangement to provide vocational training to at least one member of each displaced/other family so as to equip him/her to start his/her own small enterprise and refine his/her skills to take advantage of new job opportunities. For those engaged in traditional occupations/handicrafts/handlooms, suitable training shall be organized at the cost of project authority to upgrade their existing skills.

(c) **Convertible Preference Share:** At the option of the displaced family and, subject to the provisions of relevant law(s) in force for the time being, the project authority may issue Convertible Preference Share(s) or Secured Bond(s) up to a maximum of 50% out of one time cash assistance as mentioned in sub-Para. (a) above.

(d) **Provision for homestead land:** Subject to availability, each displaced family will be given at least 1/10th of an acre of land free of cost in a resettlement habitat for homestead purpose.

(e) **Assistance for Self-relocation:** Each of the displaced family who opts for self relocation elsewhere other than the Resettlement habitat shall be given a one time cash grant of Rs. 50, 000 in lieu of homestead land.

(f) **House Building Assistance**: Besides, Project Authority shall construct house for each of the displaced families in the resettlement habitat or provide house building assistance of Rs. 1,50,000 to each of the displaced families settling in the Resettlement habitat or opting for self-relocation elsewhere.

(g) **Shops and Service Units**: Project Authorities will also construct shops and service units at feasible locations at their own cost, which will be allotted in consultation with Collector to project displaced families opting for self-employment. While allotting such units, preference will be given to physically challenged persons and members of displaced SC & ST families.

II. Type B: Mining Projects

(a) **Employment:** Displaced and other affected families shall be eligible for employment, by the project causing displacement. For the purpose of employment, each family will nominate one member of the family. The project proponent will give preference to the nominated members of the displaced and other affected families in the matter of employment. The order of preference will be as follows:

(i) Displaced families losing all land including homestead land,
(ii) Displaced families losing more than 2/3rd of agricultural land and homestead land,
(iii) Families losing all agricultural land but not homestead land,
(iv) Displaced families losing more than 1/3rd of agricultural land and homestead land,
(v) Displaced families losing only homestead land but not agricultural land,
(vi) Families losing agricultural land in part but not homestead land.

The Project Authority will make special efforts to facilitate skill up-gradation of the nominated member of the displaced family to make him/her 'employable' in their project.

1. In case of nominees of displaced families eligible for employment otherwise, the upper age limit shall be relaxed by five years.
2. Project Authorities should notify their employment capacity sufficiently in advance.
3. As far as practicable, the objective shall be to provide one member from each displaced/other family as mentioned above with employment in the project. However, where the same cannot be provided because of reason to be explained in writing, cash compensation as mentioned below shall be provided to such families. Families, who do not opt for employment/self-employment as mentioned in sub-Para. (a) and (b), shall be provided by the Project Authority with one time

cash assistance in lieu of employment at the scale indicated below:

Sl. No.	Families under category as per sub-Para. (a) above	Amount of one time cash assistance (Rs. in lakhs)
(i)	Displaced Families coming under category (i)	5.00
(ii)	Displaced Families coming under category (ii)	3.00
(iii)	Displaced Families coming under category (iii)	2.00
(iv)	Displaced Families coming under category (iv), (v) and (vi)	1.00

(b) **Training for Self-employment:** Project Authority under the guidance of the Collector concerned will make adequate arrangement to provide vocational training to at least one member of each displaced/ other family so as to equip him/her to start his/her own small enterprise and refine his/her skills to take advantage of new job opportunities. For those engaged in traditional occupations/handicrafts/ handlooms, suitable training shall be organized at the cost of project authority to upgrade their existing skills.

(c) **Convertible Preference Share:** At the option of the displaced family the project authority may issue convertible preference share up to a maximum of 50% out of the one-time cash assistance as mentioned in sub-Para. (a) above.

(d) **Provision for homestead land:** Subject to availability, each displaced family will be given at least 1/10th of an acre of land free of cost in a resettlement habitat for homestead purpose.

(e) **Assistance for Self-relocation:** Each of the displaced family who opts for self-relocation elsewhere other than the Resettlement habitat shall be given a one time cash grant of Rs. 50,000 in lieu of homestead land.

(f) **House Building Assistance**: Besides, Project Authority shall construct house for each displaced families in the resettlement habitat or provide house building assistance of Rs. 1,50,000 to each of the displaced family settling in the Resettlement habitat or opting for self-relocation elsewhere.

(g) **Shops and Service Units**: Project Authorities will also construct shops and service units at feasible locations at their own cost, which will be allotted in consultation with Collector to project displaced families opting for self-employment. While allotting such units, preference will be given to physically challenged persons and members of displaced SC & ST families.

III. Type C: Water Resources/National Parks and Sanctuary

Rehabilitation Package for Displaced family (DF):

(a) **Homestead land** @ 0.10 acre in Rehabilitation habitat or cash equivalent @ Rs. 50,000 to each displaced family opting for self-relocation elsewhere.

(b) **House Building Assistance:** Rs. 1,50,000 to each displaced family, which includes cattle shed. This will be admissible to all displaced families whether settling in a Resettlement habitat or elsewhere.

(c) **Assistance for Agricultural Land:** Each Displaced Family belonging to ST category shall be provided two-and-a-half acres of irrigated agricultural land, or five acres of non-irrigated agricultural land.
Each Displaced Family belonging to all other categories shall be provided two acres of irrigated agricultural land, or four acres of non-irrigated agricultural land.
In case of non-availability of land, cash equivalent will be provided @ of Rs. 1,00,000 per acre of irrigated land and Rs. 50,000 per acre of non-irrigated land, including the cost of reclamation or at the rate decided by the Government from time to time.

(d) **Registration cost** of land up to 5 acres of un-irrigated land or two-and-half acres of irrigated land in case of indigenous households or 4 acres of un-irrigated land or 2 acres of irrigated land for other displaced households who opt for cash based rehabilitation in lieu of land shall be paid by the project authority within a period not exceeding five years from the date of displacement.

IV. *Type D: Urban Projects and Linear Projects*

Each displaced Family will get:

(a) **Homestead land** @ 1/10th of an acre in rural area and @ 1/25th of an acre in urban area or cash equivalent of Rs.50,000 preferably near growth centers like land by the side of roads and important junctions, land by the side of railway stations, etc., subject to availability. If required, project authority may acquire such suitable land under the relevant Act for the purpose.

(b) **House Building Assistance:** Rs. 1,50,000 to each displaced family will be admissible whether settling in a Resettlement habitat or elsewhere.

(c) If house/homestead land of any landholder is acquired for linear project or if there is total displacement due to acquisition for such project, the project authority shall provide employment to one of the members of such displaced family in the project. Wherever RPDAC decides that provision of such employment is not possible, one-time cash assistance as decided by the Government will be paid by the project authority.

V. *Type E: Any other Projects*

Government may issue separate Policy guidelines for any other projects not covered above specifying R & R packages to be extended to displaced families.

10. Benefit to Landless and Homestead-less Encroachers Common to all Categories

(a) An encroacher family, who is landless as defined in the Orissa Prevention of Land Encroachment Act, 1972, and is in possession of the encroached land at least for a period of ten years continuously prior to the date of notification under relevant law(s) declaring intention of land acquisition will get *ex gratia* equal to compensation admissible under the Land Acquisition

Act, 1894 for a similar category of land to the extent of land under his/her physical possession up to a maximum of one standard acre, if the encroachment is unobjectionable. While determining the extent of land for such compensation the *rayati* land held by him/her is to be taken into account.

(b) An encroacher family, who is homesteadless as defined in the Orissa Prevention of Land Encroachment Act, 1972 and is in possession of the encroached land at least for a period of ten years continuously prior to the date of notification under relevant law(s) declaring intention of land acquisition will get *ex gratia*, equal to compensation for the similar category of homestead land, against the encroached homestead land up to a maximum of 1/10th of an acre in rural areas or 1/25th of an acre in urban areas, if the encroachment is unobjectionable. While determining the extent of land for such compensation the homestead land held by him/her is to be taken into account. The *ex gratia* will be in addition to the actual cost of structures thereon. If the encroachment is found to be objectionable, (s)he will be entitled to the cost of structure only.

11. Other Assistance Common to all Categories of Projects

(a) **Maintenance Allowance**: In order to ensure timely vacation, an allowance of Rs. 2,000 per month per displaced family shall be provided on vacation of land/house for a period of one year from the date of vacation as determined by the Collector concerned.

(b) **Assistance for Temporary Shed:** An assistance of Rs. 10,000 shall be provided to each displaced family.

(c) **Transportation Allowance**: Transportation allowance of Rs. 2,000 or free transportation to the resettlement habitat or their new place of inhabitance, shall be provided to each displaced family by the Project Authority. Provided that State Government in Revenue Department shall review and may revise the rate if necessary, once every two years basing on the index point.

12. Additional Provisions for Assistance

Notwithstanding anything contained elsewhere in the Policy, the Government or the Project Authority may extend any additional benefits and provisions to the displaced families keeping in view the specific nature of displacement.

13. Special Benefits to Displaced Indigenous Families and Primitive Tribal Groups

(a) While developing the resettlement plans, the socio-cultural norms of indigenous and primitive tribal groups will be respected.
(b) Each displaced family of indigenous category shall be given preferential allotment of land.
(c) As far as practicable, indigenous communities should be resettled in a compact area close to their natural habitat.
(d) Indigenous displaced families resettled outside the district shall be given 25 per cent higher R & R benefits in monetary terms.

14. Indexation of Rehabilitation Grant

Rehabilitation grant will be indexed to the Wholesale Price Index (WPI) with 1-4-2006 as the reference date and will be revised by the Government once in every two years thereafter on the basis of WPI.

15. Periphery Development

The Project Authorities shall be responsible for periphery development as decided by the RPDAC within the guidelines issued from time to time by the State Government.

16. Compensation Advisory Committee

Government in Revenue Department may constitute a District Compensation Advisory Committee (DCAC) under the chairmanship of the Collector to determine negotiated price. Adequate representation will be given to women and indigenous communities (wherever applicable) in the Committee.

If any dispute arises on recommendation of the DCAC, the matter will be referred to the State Level Compensation Advisory Committee (SCAC) chaired by the Member, Board of Revenue whose decisions shall be final and binding on all concerned. The composition of this State Level Compensation Advisory Committee will be notified by the Government.

17. Rehabilitation and Periphery Development Advisory Committee (RPDAC)

In order to encourage participation of displaced people and their elected representatives in implementation and monitoring of R&R package, to oversee and monitor periphery development, the Government may constitute a Rehabilitation-*cum*-Periphery Development Advisory Committee (RPDAC) for each or a group of projects falling in one district. The detailed composition of the Committee shall be notified by the Government and it may include people's representatives, one or two leading NGOs of the affected area and select Government officers, and any other persons to be notified by the Government. Adequate representation will be given to women and indigenous communities (wherever applicable) in the Committee. Chairman of the Committee will be at liberty to co-opt members for efficient discharge of its functions.

18. State Level Council on Resettlement and Rehabilitation (SLCRR)

At the State level, there will be a Council headed by the Chief Minister to advise, review and monitor implementation of Resettlement and Rehabilitation Policy. The Council may comprise of Ministers, select representatives of the people, leading social activists, academicians and experts of national and international repute and senior officials of the Government.

19. LA and R&R Structure

(a) State Level

A Directorate of R&R will be constituted to discharge and oversee the implementation of the R&R Policy. Detailed structure and functions of this Directorate will be notified by the Government.

(b) District / Project Level

(i) Depending on magnitude of the RR works at the project level, there may be a Project Director, Resettlement and Rehabilitation (PD-RR), who shall be assisted by such other officers as Government may decide. The PD-RR shall be the Chief Coordinating Officer between all the line departments in the matter of resettlement and rehabilitation. He will be responsible for implementation of the Policy and timely execution of R&R works under the overall supervision and guidance of the Collector and the Directorate of R&R.

(ii) Where R&R work is of lesser magnitude, the same can be undertaken by the LAO of the Collectorate.

20. Budgetary Provision and Allotment

Wherever required, adequate budget provision will be kept towards land acquisition and R&R establishments to ensure effective implementation of RR Policy. The Directorate of R&R should submit such budgetary estimate/works by 1st January of every year. Government will ensure timely release of allotment to the concerned authorities.

21. Grievance Redressal Mechanism

An effective Grievance Redressal Mechanism will be set up at District and Directorate level to deal with grievance of the project displaced people relating to land acquisition, resettlement and rehabilitation. Besides, all the project authorities shall be asked to set up an effective Grievance

Redressal Mechanism relating to their project. Effective participation of the displaced communities will be ensured in the process.

22. Assessment of Policy Implementation

Government may from time to time undertake assessment of the implementation of this Policy through appropriate agencies.

23. Interpretation and Amendment:

(a) Any issues or doubts regarding this Policy shall be referred to Government in Revenue Department whose decision shall be final and binding on all concerned.

(b) Government in Revenue Department may from time to time amend the provisions as contained in this Policy as considered necessary.

(c) Government in Revenue Department shall have the powers to issue guidelines and instructions from time to time to operationalise this Policy.

Appendix-IV

The Scheduled Tribes and Other Traditional Forest Dwellers (Recognition of Forest Rights) Act, 2006

NO. 2 OF 2007
[29th December, 2006]

An Act to recognise and vest the forest rights and occupation in forest land in forest dwelling Scheduled Tribes and other traditional forest dwellers who have been residing in such forests for generations but whose rights could not be recorded; to provide for a framework for recording the forest rights so vested and the nature of evidence required for such recognition and vesting in respect of forest land.

Whereas the recognised rights of the forest dwelling Scheduled Tribes and other traditional forest dwellers include the responsibilities and authority for sustainable use, conservation of biodiversity and maintenance of ecological balance and thereby strengthening the conservation regime of the forests while ensuring livelihood and food security of the forest dwelling Scheduled Tribes and other traditional forest dwellers;

And whereas the forest rights on ancestral lands and their habitat were not adequately recognised in the consolidation of State forests during the colonial period as well as in independent India resulting in historical injustice to the forest dwelling Scheduled Tribes and other traditional forest dwellers who are integral to the very survival and sustainability of the forest ecosystem;

And whereas it has become necessary to address the long standing insecurity of tenurial and access rights of forest dwelling Scheduled Tribes and other traditional forest dwellers including those who were forced to relocate their dwelling due to State development interventions.

BE it enacted by Parliament in the Fifty-seventh Year of the Republic of India as follows:—

CHAPTER 1
PRELIMINARY

1. Short Title and Commencement

(1) This Act may be called the Scheduled Tribes and Other Traditional Forest Dwellers (Recognition of Forest Rights) Act, 2006.
(2) It extends to the whole of India except the State of Jammu and Kashmir.
(3) It shall come into force on such date as the Central Government may, by notification in the Official Gazette, appoint.

2. Definitions.— In this Act, unless the context otherwise requires,—

(a) "community forest resource" means customary common forest land within the traditional or customary boundaries of the village or seasonal use of landscape in the case of pastoral communities, including reserved forests, protected forests and protected areas such as Sanctuaries and National Parks to which the community had traditional access;
(b) "critical wildlife habitat" means such areas of National Parks and Sanctuaries where it has been specifically and clearly established, case by case, on the basis of scientific and objective criteria, that such areas are required to be kept as inviolate for the purposes of wildlife conservation as may be determined and notified by the Central Government in the Ministry of Environment and Forests after open process of consultation by an Expert Committee, which includes experts from the locality appointed by that Government wherein a representative of the Ministry

of Tribal Affairs shall also be included, in determining such areas according to the procedural requirements arising from sub-sections (1) and (2) of section 4;

(c) "forest dwelling Scheduled Tribes" means the members or community of the Scheduled Tribes who primarily reside in and who depend on the forests or forest lands for *bona fide* livelihood needs and includes the Scheduled Tribe pastoralist communities;

(d) "forest land" means land of any description falling within any forest area and includes unclassified forests, undemarcated forests, existing or deemed forests, protected forests, reserved forests, Sanctuaries and National Parks;

(e) "forest rights" means the forest rights referred to in section 3;

(f) "forest villages" means the settlements which have been established inside the forests by the forest department of any State Government for forestry operations or which were converted into forest villages through the forest reservation process and includes forest settlement villages, fixed demand holdings, all types of taungya settlements, by whatever name called, for such villages and includes lands for cultivation and other uses permitted by the Government;

(g) "Gram Sabha" means a village assembly which shall consist of all adult members of a village and in case of States having no Panchayats, Padas, Tolas and other traditional village institutions and elected village committees, with full and unrestricted participation of women;

(h) "habitat" includes the area comprising the customary habitat and such other habitats in reserved forests and protected forests of primitive tribal groups and pre-agricultural communities and other forest dwelling Scheduled Tribes;

(i) "minor forest produce" includes all non-timber forest produce of plant origin including bamboo, brush wood, stumps, cane, tussar, cocoons, honey, wax, lac, tendu or kendu leaves, medicinal plants and herbs, roots, tubers and the like;

(j) "nodal agency" means the nodal agency specified in section 11;

(k) "notification" means a notification published in the Official Gazette;

(l) "prescribed" means prescribed by rules made under this Act;

(m) "Scheduled Areas" means the Scheduled Areas referred to in clause (1) of article 244 of the Constitution;

(n) "sustainable use" shall have the same meaning as assigned to it in clause (o) of section 2 of the Biological Diversity Act, 2002;

(o) "other traditional forest dweller" means any member or community who has for at least three generations prior to the 13th day of December, 2005 primarily resided in and who depend on the forest or forests land for *bona fide* livelihood needs.

Explanation.—For the purpose of this clause, "generation" means a period comprising of twenty-five years;

(p) "village" means—

(i) a village referred to in clause (b) of section 4 of the Provisions of the Panchayats (Extension to the Scheduled Areas) Act, 1996; or

(ii) any area referred to as a village in any State law relating to Panchayats other than the Scheduled Areas; or

(iii) forest villages, old habitation or settlements and unsurveyed villages, whether notified as village or not; or

(iv) in the case of States where there are no Panchayats, the traditional village, by whatever name called;

(q) "wild animal" means any species of animal specified in Schedules I to IV of the Wild Life (Protection) Act, 1972 and found wild in nature.

CHAPTER II
FOREST RIGHTS

Forest rights of Forest dwelling Scheduled Tribes and other tradional forest dwellers.

(1) For the purposes of this Act, the following rights, which secure individual or community tenure or both, shall be the forest rights of forest dwelling Scheduled Tribes and other traditional forest dwellers on all forest lands, namely :—

(a) right to hold and live in the forest land under the individual or common occupation for habitation or for self-cultivation for livelihood by a member or members of a forest dwelling Scheduled Tribe or other traditional forest dwellers;

(b) community rights such as nistar, by whatever name called, including those used in erstwhile Princely States, Zamindari or such intermediary regimes;

(c) right of ownership, access to collect, use, and dispose of minor forest produce which has been traditionally collected within or outside village boundaries;

(d) other community rights of uses or entitlements such as fish and other products of water bodies, grazing (both settled or transhumant) and traditional seasonal resource access of nomadic or pastoralist communities;

(e) rights including community tenures of habitat and habitation for primitive tribal groups and pre-agricultural communities;

(f) rights in or over disputed lands under any nomenclature in any State where claims are disputed;

(g) rights for conversion of *Pattas* or leases or grants issued by any local authority or any State Government on forest lands to titles;

(h) rights of settlement and conversion of all forest villages, old habitation, unsurveyed villages and other villages in forests, whether recorded, notified or not into revenue villages;

(i) right to protect, regenerate or conserve or manage any community forest resource which they have been traditionally protecting and conserving for sustainable use;

(j) rights which are recognised under any State law or laws of any Autonomous District Council or Autonomous Regional Council or which are accepted as rights of tribals under any traditional or customary law of the concerned tribes of any State;

(k) right of access to biodiversity and community right to intellectual property and traditional knowledge related to biodiversity and cultural diversity;

(l) any other traditional right customarily enjoyed by the forest dwelling Scheduled Tribes or other traditional forest dwellers, as the case may be, which are not mentioned in clauses (a) to (k) but excluding the traditional right of hunting or trapping or extracting a part of the body of any species of wild animal;

(m) right to *in situ* rehabilitation including alternative land in cases where the Scheduled Tribes and other traditional forest dwellers have been illegally evicted or displaced from forest land of any description without receiving their legal entitlement to rehabilitation prior to the 13th day of December, 2005.

(2) Notwithstanding anything contained in the Forest (Conservation) Act, 1980, the Central Government shall provide for diversion of forest land for the following facilities managed by the Government which involve felling of trees not exceeding seventy-five trees per hectare, nemely:—

(a) schools;
(b) dispensary or hospital;
(c) anganwadis;
(d) fair price shops;
(e) electric and telecommunication lines;
(f) tanks and other minor water bodies;
(g) drinking water supply and water pipelines;
(h) water or rainwater harvesting structures;
(i) minor irrigation canals;
(j) non-conventional source of energy;
(k) skill upgradation or vocational training centres;

(l) roads; and
(m) community centres:

Provided that such diversion of forest land shall be allowed only if,—

(i) the forest land to be diverted for the purposes mentioned in this sub-section is less than one hectare in each case; and
(ii) the clearance of such developmental projects shall be subject to the condition that the same is recommended by the Gram Sabha.

CHAPTER III
RECOGNITION, RESTORATION AND VESTING OF FOREST RIGHTS AND RELATED MATTERS

Recognition of, and vesting of, forest rights in forest dwelling Scheduled Tribes and other traditional forest dwellers.

(1) Notwithstanding anything contained in any other law for the time being in force, and subject to the provisions of this Act, the Central Government hereby recognises and vests forest rights in—
 (a) the forest dwelling Scheduled Tribes in States or areas in States where they are declared as Scheduled Tribes in respect of all forest rights mentioned in section 3;
 (b) the other traditional forest dwellers in respect of all forest rights mentioned in section 3.
(2) The forest rights recognised under this Act in critical wildlife habitats of National Parks and Sanctuaries may subsequently be modified or resettled, provided that no forest rights holders shall be resettled or have their rights in any manner affected for the purposes of creating inviolate areas for wildlife conservation except in case all the following conditions are satisfied, namely :—
 (a) the process of recognition and vesting of rights as specified in section 6 is complete in all the areas under consideration;

(b) it has been established by the concerned agencies of the State Government, in exercise of their powers under the Wild Life (Protection) Act, 1972 that the activities or impact of the presence of holders of rights upon wild animals is sufficient to cause irreversible damage and threaten the existence of said species and their habitat;

(c) the State Government has concluded that other reasonable options, such as, co-existence are not available;

(d) a resettlement or alternatives package has been prepared and communicated that provides a secure livelihood for the affected individuals and communities and fulfils the requirements of such affected individuals and communities given in the relevant laws and the policy of the Central Government;

(e) the free informed consent of the Gram Sabhas in the areas concerned to the proposed resettlement and to the package has been obtained in writing;

(f) no resettlement shall take place until facilities and land allocation at the resettlement location are complete as per the promised package:

Provided that the critical wildlife habitats from which rights holders are thus relocated for purposes of wildlife conservation shall not be subsequently diverted by the State Government or the Central Government or any other entity for other uses.

(3) The recognition and vesting of forest rights under this Act to the forest dwelling Scheduled Tribes and to other traditional forest dwellers in relation to any State or Union territory in respect of forest land and their habitat shall be subject to the condition that such Scheduled Tribes or tribal communities or other traditional forest dwellers had occupied forest land before the 13th day of December, 2005.

(4) A right conferred by sub-section (1) shall be heritable but not alienable or transferable and shall be registered jointly in the name of both the spouses in case of

married persons and in the name of the single head in the case of a household headed by a single person and in the absence of a direct heir, the heritable right shall pass on to the next-of-kin.

(5) Save as otherwise provided, no member of a forest dwelling Scheduled Tribe or other traditional forest dweller shall be evicted or removed from forest land under his occupation till the recognition and verification procedure is complete.

(6) Where the forest rights recognised and vested by sub-section (1) are in respect of land mentioned in clause (a) of sub-section (1) of section 3 such land shall be under the occupation of an individual or family or community on the date of commencement of this Act and shall be restricted to the area under actual occupation and shall in no case exceed an area of four hectares.

(7) The forest rights shall be conferred free of all encumbrances and procedural requirements, including clearance under the Forest (Conservation) Act, 1980, requirement of paying the 'net present value' and 'compensatory afforestation' for diversion of forest land, except those specified in this Act.

(8) The forest rights recognised and vested under this Act shall include the right of land to forest dwelling Scheduled Tribes and other traditional forest dwellers who can establish that they were displaced from their dwelling and cultivation without land compensation due to State development interventions, and where the land has not been used for the purpose for which it was acquired within five years of the said acquisition.

5. Duties of holders of forest rights—The holders of any forest right, Gram Sabha and village level institutions in areas where there are holders of any forest right under this Act are empowered to—

   (a) protect the wildlife, forest and biodiversity;

   (b) ensure that adjoining catchments area, water sources and other ecological sensitive areas are adequately protected;

(c) ensure that the habitat of forest dwelling Scheduled Tribes and other traditional forest dwellers is preserved from any form of destructive practices affecting their cultural and natural heritage;

(d) ensure that the decisions taken in the Gram Sabha to regulate access to community forest resources and stop any activity which adversely affects the wild animals, forest and the biodiversity are complied with.

CHAPTER IV
AUTHORITIES AND PROCEDURE FOR VESTING OF FOREST RIGHTS

Authorities to vest forest rights in forest dwelling Scheduled Tribes and other traditional forest dwellers and procedure thereof.

6. (1) The Gram Sabha shall be the authority to initiate the process for determining the nature and extent of individual or community forest rights or both that may be given to the forest dwelling Scheduled Tribes and other traditional forest dwellers within the local limits of its jurisdiction under this Act by receiving claims, consolidating and verifying them and preparing a map delineating the area of each recommended claim in such manner as may be prescribed for exercise of such rights and the Gram Sabha shall, then, pass a resolution to that effect and thereafter forward a copy of the same to the Sub-Divisional Level Committee.

(2) Any person aggrieved by the resolution of the Gram Sabha may prefer a petition to the Sub-Divisional Level Committee constituted under sub-section (3) and the Sub-Divisional Level Committee shall consider and dispose of such petition:

Provided that every such petition shall be preferred within sixty days from the date of passing of the resolution by the Gram Sabha:

Provided further that no such petition shall be disposed of against the aggrieved person, unless he has been given a reasonable opportunity to present his case.

(3) The State Government shall constitute a Sub-Divisional Level Committee to examine the resolution passed by the Gram Sabha and prepare the record of forest rights and forward it through the Sub-Divisional Officer to the District Level Committee for a final decision.

(4) Any person aggrieved by the decision of the Sub-Divisional Level Committee may prefer a petition to the District Level Committee within sixty days from the date of decision of the Sub-Divisional Level Committee and the District Level Committee shall consider and dispose of such petition:

Provided that no petition shall be preferred directly before the District Level Committee against the resolution of the Gram Sabha unless the same has been preferred before and considered by the Sub-Divisional Level Committee:

Provided further that no such petition shall be disposed of against the aggrieved person, unless he has been given a reasonable opportunity to present his case.

(5) The State Government shall constitute a District Level Committee to consider and finally approve the record of forest rights prepared by the Sub-Divisional Level Committee.

(6) The decision of the District Level Committee on the record of forest rights shall be final and binding.

(7) The State Government shall constitute a State Level Monitoring Committee to monitor the process of recognition and vesting of forest rights and to submit to the nodal agency such returns and reports as may be called for by that agency.

(8) The Sub-Divisional Level Committee, the District Level Committee and the State Level Monitoring Committee shall consist of officers of the

departments of Revenue, Forest and Tribal Affairs of the State Government and three members of the Panchayati Raj Institutions at the appropriate level, appointed by the respective Panchayati Raj Institutions, of whom two shall be the Scheduled Tribe members and at least one shall be a woman, as may be prescribed.

(9) The composition and functions of the Sub-Divisional Level Committee, the District Level Committee and the State Level Monitoring Committee and the procedure to be followed by them in the discharge of their functions shall be such as may be prescribed.

CHAPTER V
OFFENCES AND PENALTIES

Offences by members or officers of authorities and Committees under this Act.

7. Where any authority or Committee or officer or member of such authority or Committee contravenes any provision of this Act or any rule made thereunder concerning recognition of forest rights, it, or they, shall be deemed to be guilty of an offence under this Act and shall be liable to be proceeded against and punished with fine which may extend to one thousand rupees:

   Provided that nothing contained in this sub-section shall render any member of the authority or Committee or head of the department or any person referred to in this section liable to any punishment if he proves that the offence was committed without his knowledge or that he had exercised all due diligence to prevent the commission of such offence.

8. Cognizance of offences.—No court shall take cognizance of any offence under section 7 unless any forest dwelling Scheduled Tribe in case of a dispute relating to a resolution of a Gram Sabha or the Gram Sabha through a resolution against any higher

authority gives a notice of not less than sixty days to the State Level Monitoring Committee and the State Level Monitoring Committee has not proceeded against such authority.

CHAPTER VI
MISCELLANEOUS

Members of authorities, etc., to be public servants.

9. Every member of the authorities referred to in Chapter IV and every other officer exercising any of the powers conferred by or under this Act shall be deemed to be a public servant within the meaning of section 21 of the Indian Penal Code.
10. Protection of action taken in good faith.—
    (1) No suit, prosecution or other legal proceeding shall lie against any officer or other employee of the Central Government or the State Government for anything which is in good faith done or intended to be done by or under this Act.
    (2) No suit or other legal proceeding shall lie against the Central Government or the State Government or any of its officers or other employees for any damage caused or likely to be caused by anything which is in good faith done or intended to be done under this Act.
    (3) No suit or other legal proceeding shall lie against any authority as referred to in Chapter IV including its Chairperson, members, member-secretary, officers and other employees for anything which is in good faith done or intended to be done under this Act.
11. Nodal agency.—The Ministry of the Central Government dealing with Tribal Affairs or any officer or authority authorised by the Central Government in this behalf shall be the nodal agency for the implementation of the provisions of this Act.
12. Power of Central Government to issue directions.—In the performance of its duties and exercise of its powers

by or under this Act, every authority referred to in Chapter IV shall be subject to such general or special directions, as the Central Government may, from time to time, give in writing.

13. Act not in derogation of any other law.—Save as otherwise provided in this Act and the Provisions of the Panchayats (Extension to the Scheduled Areas) Act, 1996, the provisions of this Act shall be in addition to and not in derogation of the provisions of any other law for the time being in force.

14. Power to make rules.—

(1) The Central Government may, by notification, and subject to the condition of previous publication, make rules for carrying out the provisions of this Act.

(2) In particular, and without prejudice to the generality of the foregoing powers, such rules may provide for all or any of the following matters, namely:—

(a) procedural details for implementation of the procedure specified in section 6;

(b) the procedure for receiving claims, consolidating and verifying them and preparing a map delineating the area of each recommended claim for exercise of forest rights under sub-section (1) of section 6 and the manner of preferring a petition to the Sub-Divisional Committee under sub-section (2) of that section;

(c) the level of officers of the departments of Revenue, Forest and Tribal Affairs of the State Government to be appointed as members of the Sub-Divisional Level Committee, the District Level Committee and the State Level Monitoring Committee under sub-section (8) of section 6;

(d) the composition and functions of the Sub-Divisional Level Committee, the District Level Committee and the State Level Monitoring Committee and the procedure to be followed by them in the discharge of their functions under sub-section (9) of section 6;

(e) any other matter which is required to be, or may be, prescribed.

(3) Every rule made by the Central Government under this Act shall be laid, as soon as may be after it is made, before each House of Parliament, while it is in session, for a total period of thirty days which may be comprised in one session or in two or more successive sessions, and if, before the expiry of the session immediately following the session or the successive sessions aforesaid, both Houses agree in making any modification in the rule or both Houses agree that the rule should not be made, the rule shall thereafter have effect only in such modified form or be of no effect, as the case may be; so, however, that any such modification or annulment shall be without prejudice to the validity of anything previously done under that rule.

Appendix-V

Household Questionnaire Schedule

A. HOUSEHOLD CHARACTERISTICS

1. Name of the District-
2. Name of the Block-
3. Name of the GP-
4. Name of the Village-
5. Name of the Respondent-
6. Name of the head of the Household-
7. Religion of the Head-
   Hindu-1, Christian-2, Muslim-3, other-4
8. Since how many generations he has been in this religion
9. Caste of the head-
   SC-1, ST-2, OBC-3, General-4
10. If belongs to SC/ST, what is your sub-caste.
11. Ownership of House-
    1- Own, 2- Rent, 3- Adjust, 4 - Any Other
12. Family type-
    1-joint, 2-nuclear

Basic Household Information

Sl. no	*Name of HH members (start with name of HHH)*	*Relation with HHH(a)*	*Age*	*Sex (b)*	*Marital status (c)*	*Qualifi-cation(d)*	*Occupa-tion (e)*	*Annual Income (f)*
1	2	3	4	5	6	7	8	9
1								
2								
3								

Sl. no	*Name of HH members (start with name of HHH)*	*Relation with HHH(a)*	*Age*	*Sex (b)*	*Marital status (c)*	*Qualifi-cation(d)*	*Occupa-tion (e)*	*Annual Income (f)*
1	*2*	*3*	*4*	*5*	*6*	*7*	*8*	*9*
4								
5								
6								
7								
8								
9								
10								

Code-

(a) 1-head/self, 2-wife, 3-husband, 4-son, 5-daughter, 6-brother, 7-sister, 8-father, 9-mother, 10–grandson, 11–granddaughter, 12-father-in-law, 13-mother-in-law, 14-Any other

(b) 1-male, 2–female

(c) 1–married, 2–unmarried

(d) 1-illiterate, 2-below primary level, 3-primary level, 4-middle school level, 5-Matriculation, 6-Graduation, 7-Post-Graduation, 8-Any other

(e) 1-small and-marginal farmer, 2-agricultural labourer, 3-horticulture, 4-animal husbandry, 5-Minor forest Produce, 6-cottage industry, 7-govt. service, 8-private service, 9-wage labourer, 10-business, 11-retired, 12--other (specify)

(f) 1-(1-10000), 2-(10001-20000), 3-(20001-30000), 4-(30001-40000), 5-(40001 and above)

Sl. no	*Heads of Expenditure*	*Annual Average expenditure*			
		0-2,000(1)	*2,001-3,000 (2)*	*3,001-4,000(3)*	*4,001 and above(4)*
1	2	3	4	5	6
1	Food				
2	Clothing				
3	Maintenance of House				
4	Health				
5	Education				

Sl. no	*Heads of Expenditure*	*Annual Average expenditure*			
		0-2,000(1)	*2,001-3,000 (2)*	*3,001-4,000(3)*	*4,001 and above(4)*
1	2	3	4	5	6
6	Alcohol / drugs				
7	Other intoxicants				
8	Social Functions/Rituals				
9	Entertainment				
10	Litigation				
11	Household durable consumer's goods				
12	Others				
Total					

B. LAND RELATED INFORMATION

1. Nature of Ownership of land
   Own Possession-1, If yes, collect quantity of all categories of land Possession through Inheritance-2, if yes, collect quantity of land except *patta* land
   No Possession – 3 (not a single type of land possessed)
2. Categorywise Land Holdings/and Ownership (at the time of Survey)

Code	*Category of Land*	*Size of Land Holding (In Acres)*			
		Patta land (1)	*Leased land (2)*	*Encroached land (3)*	*Land owned (4) (1+2+3)*
1	Dangara-I				
2	Dangara-II				
3	Dangara-III				
4	Paddy-I				
5	Paddy-II				
6	Paddy-III				
7	Atala-I				
8	Atala-II				
9	Atala-III				
10	Garden				
11	Homestead land				
12	Others				
Total					

Note: Land owned= Patta land + Leased land+ Encroached land

C. LAND ALIENATION

1. Mention whether any of the following type of land was sold, leased and mortgaged, transferred.

Nature of Transfer—(A) Sale

Code	Category of Land	Type of Transfer											
		1	2	3	4	5	6	7	8	9	10	11	12
1	Dangara-I												
2	Dangara-II												
3	Dangara-III												
4	Paddy-I												
5	Paddy-II												
6	Paddy-III												
7	Atala-I												
8	Atala-II												
9	Atala-III												
10	Garden												
11	Homestead land												
12	Others												

Total

Note: SC to SC-1, SC to ST-2, SC to Gen-3, ST to ST-4, ST to SC-5, ST to General-6, General to SC-7, General to ST-8, Others to ST-9, Transfer between any Parties-10, Land holder to company-11, Land holder to Govt-12

Nature of Transfer—(B) Mortgage

Code	Category of Land	Type of Transfer											
		1	2	3	4	5	6	7	8	9	10	11	12
1	Dangara-I												
2	Dangara-II												
3	Dangara-III												
4	Paddy-I												
5	Paddy-II												
6	Paddy-III												

Code	Category of Land	Type of Transfer											
		1	2	3	4	5	6	7	8	9	10	11	12
7	Atala-I												
8	Atala-II												
9	Atala-III												
10	Garden												
11	Homestead land												
12	Others												
Total													

Note: *SC to SC-1, SC to ST-2, SC to Gen-3, ST to ST-4, ST to SC-5, ST to General-6, General to SC-7, General to ST-8, Others to ST-9, Transfer between any Parties-10, Land holder to company-11, Land holder to Govt-12*

Nature of Transfer—(C) Transfer

Code	Category of Land	Type of Transfer											
		1	2	3	4	5	6	7	8	9	10	11	12
1	Dangara-I												
2	Dangara-II												
3	Dangara-III												
4	Paddy-I												
5	Paddy-II												
6	Paddy-III												
7	Atala-I												
8	Atala-II												
9	Atala-III												
10	Garden												
11	Homestead land												
12	Others												
Total													

Note: SC to SC-1, SC to ST-2, SC to Gen-3, ST to ST-4, ST to SC-5, ST to General-6, General to SC-7, General to ST-8, Others to ST-9, Transfer between any Parties-10, Land holder to company-11, Land holder to Govt-12

3. Causes of Land Alienation

Form of alienation	*To whom alienation name & address*	*Cause of Transfer*													
		1	2	3	4	5	6	7	8	9	10	11	12	13	14
1. Sale															
2. Mortgage															
3. Mortgage with money-lenders															
4. Benami transfer															
5. Forcible Occupation by non-tribals															
6. Oral Transfer															
7. Tribal spouse															
8. Lease															
9. Encroachment															
10. Public Purpose															
11. Industries/ Factories/Mining															
12. Voluntary Organization															
13. Transfer															
14. Land Acquisition															
15. Exchange															
16. Donation															
17. Cheating															
18. Consolidation of land holding															
19. Court Order															
20. Survey and Settlement															
21. Mutual Understanding															
22. Others															

Codes of the Land Transfer (Put Tick)

Note: *Household expenditure-1, House Construction and Repairing-2, Repayment of debt-3, Social and Religious functions-4, Agricultural Improvement-5, Bullock purchase-6, Purchase of inputs/raw-materials-7, Litigation (Court cases)-8, Medical Treatment- 9, Alchohol/Drugs-10, Natural Calamities (Drought, Floods etc)-11, Unable to Cultivate-12, To meet loss occurred due to crop failure/ Productivity/non-profit enterprises- 13, Land Acquisition-14*

4. *Whether the Respondent is Aware of any Legislation Prohibiting Transfer of Land*

Yes-1, No-2

If yes, which of the following Acts/Laws

1. OLR-1,
2. PESA-2,
3. Regulation (II) of 1956-3,
4. Orissa Land Acquisition Act (1986)-4,
5. Orissa Survey and Settlement Act (1958)-5,
6. Orissa Land Passbook Law-6,
7. Mo Jami Mo Diha-7,
8. Others-8

D. RESTORATION OF LAND

1. Have you ever (the respondent) approached any Revenue/Govt. official/Competent Authority for restoration of alienated land. *Yes-1, No-2*
   (i) If yes, how could you know the same?
   Own initiative-1
   Motivated by some body else-2
   Media-3
   NGOs-4
   Others-5
   (ii) Mention if there is any positive result:
   Yes-1, No-2

If yes, what and how, please explain.

2. Whether any case has been instituted in the court of law to regain the possession of alienated land.

Yes–1 , No–2

If yes, give details
What is the result after case(s)
Situation remains as before-1
Land has been restored to him-2

3. Whether land is now under his possession and he is cultivating it. *Yes-1, No-2*

4. Time taken for restoration of the alienated land
   0-6 months-1
   6-one year-2
   One year-2 years-3
   More than 2 years-4

Open-ended Questions

1. Experience and opinions of the respondent regarding methods of prevention of land alienation and its restoration.
   (Use extra sheet if required)
2. Case Studies (if Any) regarding land alienation based on the following reasons:
   Household expenditure-1,
   House Construction and Repairing-2,
   Repayment of debt-3,
   Social and Religious functions-4,
   Agricultural Improvement-5,
   Bullock purchase-6,
   Purchase of inputs/raw-materials-7.
   Litigation (Court cases)-8
   Medical Treatment-9,
   Alcohol/Drugs-10,
   Natural Calamities (Drought, Floods etc.)-11,
   Unable to Cultivate-12,
   To meet loss occurred due to crop failure/ Productivity/non-profit enterprises-13.
   Any woman suffered due to alienation
3. Opinion of the Investigator with signature and date
   (Use extra sheet if required)
4. Opinion of the supervisor with signature and date
   (Use extra sheet if required)

Appendix-VI

Method of Land Alienation in all Households of Koraput District

Village name	Quantity of land Alienated									
	1	2	3	4	5	6	7	8	9	10
Devighat										
No. of transfers	5	0	0	0	0	2	0	0	0	0
Extent of area covered	4.48	0.0	0.0	0.0	0.0	2.14	0.0	0.0	0.0	0.0
Dhaudapadar										
No. of transfers	2	6	0	2	0	0	0	0	1	0
Extent of area covered	1.97	12.46	0.0	4.05	0.0	0.0	0.0	0.0	18	0.0
Machhput										
No. of transfers	0	3	0	0	0	0	0	1	13	0
Extent of area covered	0.0	4.17	0.0	0.0	0.0	0.0	0.0	4	119.82	0.0
Nighamaniguda										
No. of transfers	2	5	0	1	0	2	0	0	21	0
Extent of area covered	2	42.99	0.0	2.2	0.0	11	0.0	0.0	110.09	0.0

Contd.

Village name	*Quantity of land Alienated*									
	1	*2*	*3*	*4*	*5*	*6*	*7*	*8*	*9*	*10*
Bhitaragada										
No. of transfers	1	0	0	0	0	0	0	1	1	0
Extent of area covered	1	0.0	0.0	0.0	0.0	0.0	0.0	0.7	0.5	0.0
Panchada										
No. of transfers	13	8	1	0	0	5	0	0	0	0
Extent of area covered	21.71	13.28	1.2	0.0	0.0	6.57	0.0	0.0	0.0	0.0
Mukhibedei										
No. of transfers	3	29	0	1	0	0	0	0	1	1
Extent of area covered	6.48	60.75	0.0	2	0.0	0.0	0.0	0.0	0.2	8.86
Sriramput										
No. of transfers	3	3	0	0	0	3	2	0	5	6
Extent of area covered	2.9	2.95	0.0	0.0	0.0	23.98	2.57	0.0	67.76	23.91
Mania										
No. of transfers	20	14	0	0	1	4	4	2	0	1
Extent of area covered	60.06	34.32	0.0	0.0	3.5	15.44	3.86	3.5	0.0	3.65
Aligan										
No. of transfers	6	10	0	0	0	0	0	0	0	0
Extent of area covered	18.46	71.96	0.0	0.0	0.0	0.0	0.0	0.0	0.0	0.0
Baraguda										
No. of transfers	0	0	0	0	0	0	0	0	0	0
Extent of area covered	0.0	0.0	0.0	0.0	0.0	0.0	0.0	0.0	0.0	0.0

Contd.

Village name	*Quantity of land Alienated*									
1	*2*	*3*	*4*	*5*	*6*	*7*	*8*	*9*	*10*	
Jhadipadar										
No. of transfers	0	0	0	0	0	0	0	0	0	0
Extent of area covered	0.0	0.0	0.0	0.0	0.0	0.0	0.0	0.0	0.0	0.0
Total										
No. of Transfers	55	78	1	4	1	16	6	4	42	8
Extent of area covered	119.06	242.88	1.2	8.25	3.5	59.13	6.43	8.2	316.37	36.42

Note: 1. Sale; 2. Mortgage; 3. Mortgage with money-lenders; 4. Oral Transfers; 5. Lease; 6. Encroachment; 7. Public purpose; 8. Transfer; 9. Land Acquisition; 10. Cheating

Appendix-VII

Method of Land Alienation of ST Households of Koraput District

Village name	*Quantity of land Alienated*								
	1	*2*	*3*	*4*	*5*	*6*	*7*	*8*	*9*
Devighat									
No. of transfers	3	0	0	0	1	0	0	0	0
Extent of area ccvered	3.05	0.0	0.0	0.0	0.24	0.0	0.0	0.0	0.0
Dhaudapadar									
No. of transfers	2	6	0	2	0	0	0	0	0
Extent of area covered	1.97	12.46	0.0	4.05	0.0	0.0	0.0	0.0	0.0
Machhput									
No. of transfers	0	3	0	0	0	0	1	13	0
Extent of area covered	0.0	4.17	0.0	0.0	0.0	0.0	4	119.82	0.0
Nighamaniguda									
No. of transfers	2	1	0	1	2	0	0	16	0
Extent of area covered	2	30	0.0	2.2	11	0.0	0.0	102.02	0.0

Contd.

Village name	*Quantity of land Alienated*								
	1	*2*	*3*	*4*	*5*	*6*	*7*	*8*	*9*
Panchada									
No. of transfers	3	3	1	0	3	0	0	0	0
Extent of area covered	5.03	3.05	1.2	0.0	4.04	0.0	0.0	0.0	0.0
Mukhibedei									
No. of transfers	2	25	0	1	0	0	0	1	0
Extent of area covered	1.48	53.34	0.0	2	0.0	0.0	0.0	0.2	0.0
Sriramput									
No. of transfers	2	2	0	0	2	1	0	2	2
Extent of area covered	2.5	1.95	0.0	0.0	22.97	1	0.0	3.72	4
Mania									
No. of transfers	5	2	0	0	0	0	0	0	0
Extent of area covered	13.12	0.39	0.0	0.0	0.0	0.0	0.0	0.0	0.0
Aligan									
No. of transfers	2	9	0	0	0	0	0	0	0
Extent of area covered	13.16	60.22	0.0	0.0	0.0	0.0	0.0	0.0	0.0
Bhitaragada									
No. of transfers	0	0	0	0	0	0	0	0	0
Extent of area covered	0.0	0.0	0.0	0.0	0.0	0.0	0.0	0.0	0.0
Baraguda									
No. of transfers	0	0	0	0	0	0	0	0	0
Extent of area covered	0.0	0.0	0.0	0.0	0.0	0.0	0.0	0.0	0.0

Contd.

Contd.

Village name	*Quantity of land Alienated*								
	1	*2*	*3*	*4*	*5*	*6*	*7*	*8*	*9*
Jhadipadar									
No. of transfers	0	0	0	0	0	0	0	0	0
Extent of area covered	0.0	0.0	0.0	0.0	0.0	0.0	0.0	0.0	0.0
Total									
No. of Transfers	21	51	1	4	8	1	1	32	2
Extent of area covered	42.31	165.58	1.2	8.25	38.25	1	4	225.76	4

Note: 1. Sale; 2. Mortgage; 3. Mortgage with money-lenders; 4. Oral Transfers; 5. Encroachment; 6. Public purpose; 7. Transfer; 8. Land Acquisition; 9. Cheating.

Appendix-VIII

Case Studies

During the course of interaction with the villagers, key stakeholders, households and other target groups, some important case studies were recorded. The information was cross-checked and verified from local people with government records. The following case studies are recorded in the study areas of Koraput and Rayagada districts:

KORAPUT DISTRICT

Case Study No. 1 (Forcible Occupation)

A tribal named Rama Majhi of Nigamaniguda village of Padampur GP of Koraput block alleged that he has 12.75 acres of land in the name of his forefathers. Out of these 12.75 acres of land, 9 acres were forcibly taken away by two tribal brothers. The two brothers now enjoy the produce. The issue was raised in Gram Sabha. Despite all his efforts, he failed to get his land back. ISS advised to file a case against the encroachers at Sub-collector's office.

Case Study No. 2 (Forcible Occupation)

A *dalit* of D.Panchada village of Panchada GP of Laxmipur block has been deprived of enjoying his father's 20 acres land which was forcibly taken away by two persons (one general, another OBC). The RoR (patta) of the land is in the name of his father. RoR (patta) is with the father whereas land is enjoyed by the two persons. Due to his ignorance and financial shortage he has not taken any strong step. He was advised to file petition before the Sub-collector/Tehsildar concerned.

Case Study No-3 (Cheating)

A 35 years young tribal namely Chatar Muduli of Devighat village of Devighat GP of Koraput block alleged that 30 years back when his father was alive, a general caste person motivated his father to take loan from the bank for poultry farm business. He was asked to deposit his *patta* as security for sanction of loan. At present, the general caste person is running the poultry business and has occupied the land, constructed a house over it and cultivating the land. Since *patta* is in the name of his father, the bank asked him to repay the loan. He is not in a position to repay. The most agony part is that the tribe is in a false position for no fault of him. The non-tribal person cheated him. He was advised to approach the Tehasildar/Sub-collector with adequate evidence or file a case in civil court through legal aid cell.

Case Study No. 4 (Cheating)

A cheating case was reported by one Gopi Injon.who has become a victim of Sriramput village of Rajput GP of Semiliguda block as one government official of forest department took away their lands by manipulating caste certificate without leaving any authentic evidence. At the same time, it came to the notice that the people having the surname 'Rana' of the village actually belong to scheduled tribe. But there are persons having two sets of *pattas*, one *patta* with Rana as OBC title and another Rana as ST title. Taking advantage of this anomalous position, fraudulent transaction took place between the tribal and the non-tribal keeping no written evidence as a result of which the non-tribal manipulated records and got land recorded in his name. The tribal was advised to file petition before the Tahasildar/Sub-collector for restoration of the illegally transferred land.

Case Study No. 5 (Encroachment)

A tribal old lady namely Smt. Haribala BadaNayak of Nigamaniguda village of Padampur GP of Koraput block sold one acre to a Christian Mission for constructing a school and in exchange she received Rs. 8000. In the process of construction of school, they occupied two more acres which they encroached upon. She has not been paid anything for the encroached land. Now she is undergoing severe hardships. It is very difficult on her part to meet the household expenses. She is in possession of some land but cannot cultivate due to paucity of money. Now, she is interested to sell the available land with her to meet family's day-to-day expenditure. After consulting RI regarding the encroachment by the Christian Mission, she filed a case. But she has not got any result yet. She was advised to file reminder petitions before the Tehasildar/Sub-collector and in the grievance cell of district collector for relief.

Case Study No. 6 (Cheating)

A tribal widow namely Smt. Kousalya of Rajput village of Rajput GP of Semiliguda block reported that all her immovable properties were transferred to a non-tribal when her husband was alive. By administering liquor to her husband, all her properties were snatched away by the non-tribal. At present she is landless and earns her livelihood as a wage-labour and she is maintaining her family expenses. The lady being alone and ignorant has not filed any case and also has not taken any step for restoration of the land. She was advised to file a petition before the Tehasildar/Sub-collector for redressal.

RAYAGADA DISTRICT

Case No. 1 (Household Expenditure)

A person named Nachika Basana son of Katri, a Kandha tribal, from Garadi village of Chanchuraguda Panchayat of Bisamcuttack block transferred his land by sale to a non-tribal named Kamalama Padhiani, wife of Radha Mohan Padhi, Rayagada vide MC No. 374/72 of 1.57 acres. The tribal person said he has not received any notice issued by Sub-collector for restoration. The details of the transferred land are as follows:

Khata No.	*Plot No.*	*Category*	*Acres*
11	28/1	Paddy-2	0.26
88	Do	Paddy-1	0.17
28	Do	Paddy-2	1.14

Case No. 2 (Social Functions)

A tribal in Garadi village of Chanchuraguda Panchayat of Bisamcuttack block named Deburu Adangaka son of Relli sold his land (5.90 acres) to a non-tribal to meet expenses in connection with social functions. Regulation 2/56 under section 3B (1), case no. 433/04 Sub-Collector Gunupur issued a notice for restoration fixing the hearing date to 7/4/08 in memo no. 2684 sent on 20/3/08. The tribal could not attend for hearing because he had no money even to attend the court. He was advised to meet the District Welfare Office to take help of the Legal Aid Cell and attend the court.

Case No. 3 (Cheating)

In village Nathama of Tadama GP of Rayagada block lands of some tribal households were transferred illegally. Few years back a Sundhi named Niranjan Rana was collecting revenues from innocent tribals to deposit in the revenue office in their names but cunningly he got the lands transferred to his name and subsequently deposited revenues in his own name. The villagers filed restoration case in the Sub-collector office. Case is still pending. They were advised to pursue the case and if required take help of legal cell through DWO/ADWO/WEO for redressal.

Case No. 4 (Cheating)

A tribal namely Sansa Kadraka, son of Sukru Kadraka of Gunakhal village of Therubali GP of Kolnara block aged about 70 reported that his father's elder brother Dama Kadraka mortgaged 4 acres of land with a non-tribal for Rs. 120 only. After the death of the non-tribal, his son has made *patta* in his name for those lands illegally and now refuses to give back the land to the tribal. No case has been filed for restoration. He was advised to file petition before the Sub-collector with details and contact offices of Tribal Welfare department for help through Legal Aid Cell.

Case No. 5 (Cheating)

Another tribal of Gunakhal village namely Arika Huika of Therubali GP of Kolnara block mortgaged 3 acres of land with Gasi Senapati (non-tribal) of Sintiguda few years back who made *patta* in his name illegally and further sold to another tribal in 2007. He was advised to file petition before the Sub-collector for restoration.

Bibliography

Annual Land Revenue Administration Report of Koraput 2006-07 and Rayagada District 2005-08.

Behura, N.K. &. Mohanti, K.K. (2009), "Tribe vis-à-vis land critical appraisal of issue and option in Orissa", Article published in the book "Critical Issues in Tribal Development" Edited by A.B. Ota, Published by Director SCSTRTI, Orissa.

Behura, N.K. and N. Panigrahi,(2001) A Report on Functioning of the Fifth Schedule of the constitution in the state of Orissa, Nabakrusna Chcudhury Centre for Development Studies, Bhubaneswar.

Behuria, N.C. (1965), Final Report on the Major Settlement Operation in Koraput District.

Bordoloi. B.N. (1991), Report on Transfer and Alienation of Tribal Land in Assam with special reference to the Karbis of the Karbi, Anglong district.

Dash, G. (1989) "The Land Systems and Land Reforms in Orissa", The Nabajiban Press, Cuttack.

Dash, Giridhari. (2004). Law Relating to Government Land and Lease Principles in Orissa. Cuttack,

Dash, S.N (2001), "A Study of the problem of Land Alienation in Tribal areas with special reference to the district of Koraput". Ph.D Thesis.

"Development, Displacement and Resettlement" by Manish Kumar Verma, Rawat Publication, 2004.

District Statistical Hand Book, Koraput, 2005, Directorate of Economics and Statistics Orissa, Bhubaneswar.

Economic Survey, 2006-07, Government of Orissa.

Hazzra, Amit (2006), "Land Reforms – Myths and Realities", Concept Publishing Company Pvt. Ltd., New Delhi.

Jha, Vinod (1977), "Orissa Chapter in Land Alienation and Restoration in Tribal Communities in India", Ed by Dubey, S.N. and Murdia, Himalaya Publishing House, Bombay.

Kalahandi Survey and Settlement Report, Cuttack, Orissa Government Press.

Majhi, U.N (2009), "Tribal Land Alienation: Needs Radical Solution", Article published in the book Critical Issues in Tribal Development Edited by A.B. Ota, Published by Director SCSTRTI, Orissa.

Mearns, Rabin and Saurav Sinha. (1998). Social Exclusion and Land Administration in Orissa, India. Washington, World Bank.

Mishra, G.P (1996), "Distribution of Surplus Land and Rural Poor", APH Publishing Corporation, New Delhi.

Mohapatra, M.D (2009) "The Forest Rights Act: An Instrument to ensure Forest Dwellers Rights", Article published in the book "Two Tribal Friendly Acts and Their Implications" edited by A.B. Ota & Krunakar Patanaik, Published by Director SCSTRTI, Orissa.

Mohapatro, P.C. (1987), "Economic Development of Tribal India", Ashish Publishing House, New Delhi.

Pandey, Balaji (1998), Depriving the Underprivileged for Development, Institute of Socio-Economic Development, Bhubaneswar.

Panigrahi, N (2001), Impact of State Policies on management of Land Resources in Tribal Areas of Orissa, Man & Development, March, 2001.

Patnaik, K.K. (2009), "Land Alienation and Other Land Related Issues in Tribal Areas", Article published in the book "Critical Issues in Tribal Development" Edited by A.B.Ota, Published by Director SCSTRTI, Orissa.

Patnaik, S. (2009), "Forest and tribal livelihood" Article published in the book "Two Tribal Friendly Acts and Their Implications" edited by A.B. Ota & Krunakar Patanaik, Published by Director SCSTRTI, Orissa.

Prakasham. G. (2004), "Tribals and Their Right to livelihood; Tribal Right to Land in Madhya Pradesh", Edited by Sah and Yatindra Singh Sisodia.

Raghava Rao, D.V. Indebtedness in Tribal Society, Journal of Social Research, Vol. XIV No. 1.

Report on the study of Tribal Land Alienation in Orissa, Scheduled Caste and Scheduled Tribe Research and Training Institute, Government of Orissa, Bhubaneswar (1987).

Sahoo, A.C. (2009), "The scheduled tribes and other traditional forest dwellers (recognition of Forest Rights) Act, 2006: a critical view on forest policy, practice and operational issues", Article published in the book "Two Tribal Friendly Acts and Their Implications" edited by A.B. Ota & Krunakar Patanaik, Published by Director SCSTRTI, Orissa.

Sarangi. S.P. Misra & S. Behera (2005), Research Study on 'A Socio-Economic and Legal Study of Scheduled Tribes' Land in Orissa.

Satapathy, B. (1977), Land Reform Administration in Orissa, Institute of Behavioural Sciences, Bodhagaya, India.

Singh. S.K. (Ed) (2005), A Series on Self-Governance for Tribals, National Institute of Rural Development (NIRD), Hyderabad.

Study Report on "Women and Land Alienation in Jharkhand, Chattisgarh and Orissa" (2007) Institute of Socio-Economic Development, Bhubaneswar.

"The Land Acquisition Act, 1894", (2005) Universal Law Publishing Co. Pvt. Ltd.

UNDP Status Report—Land Rights and Ownership in Tribal Areas (2008).

Viegas, P. (1991), Encroached and enslaved: alienation of tribal lands and its dynamics, New Delhi, Indian Social Institute.

Yunus, S.A. (1977), "Orissa Chapter in Land Alienation and Restoration in Tribal Communities in India", Ed by Dubey, S.N. and Murdia, Himalaya Publishing House, Bombay.

Maps

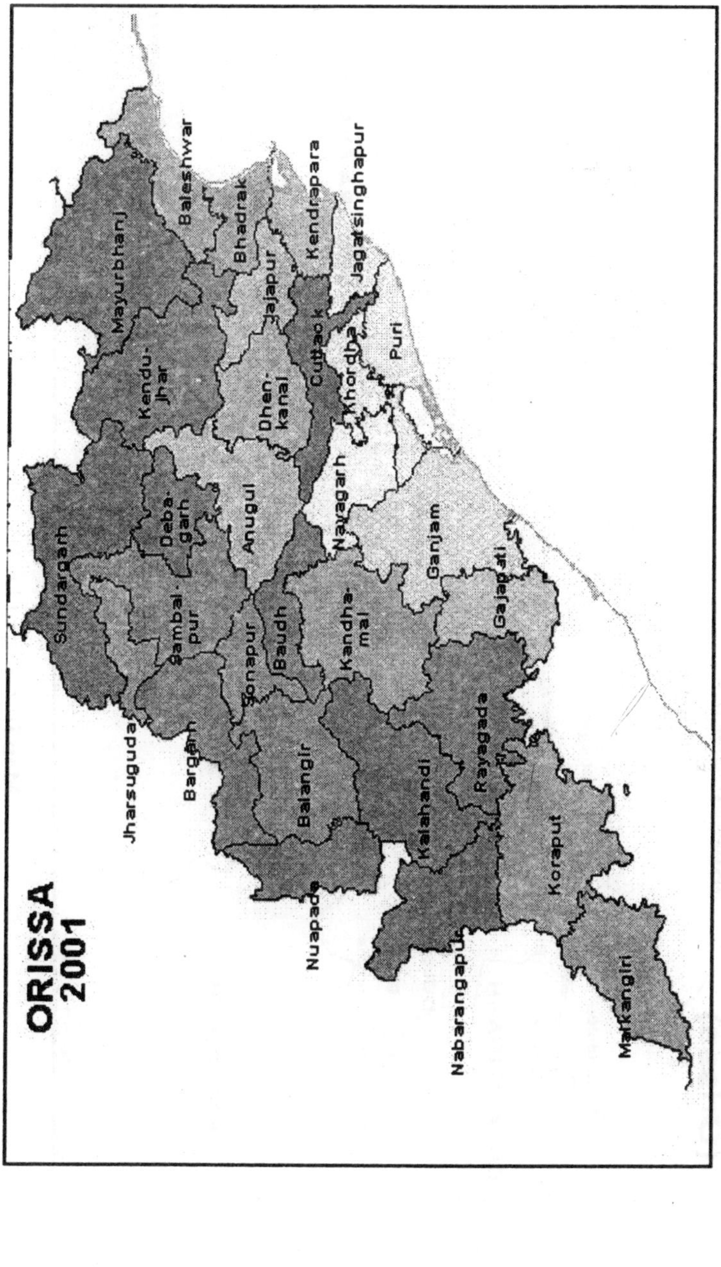

Map I: Orissa State Map.

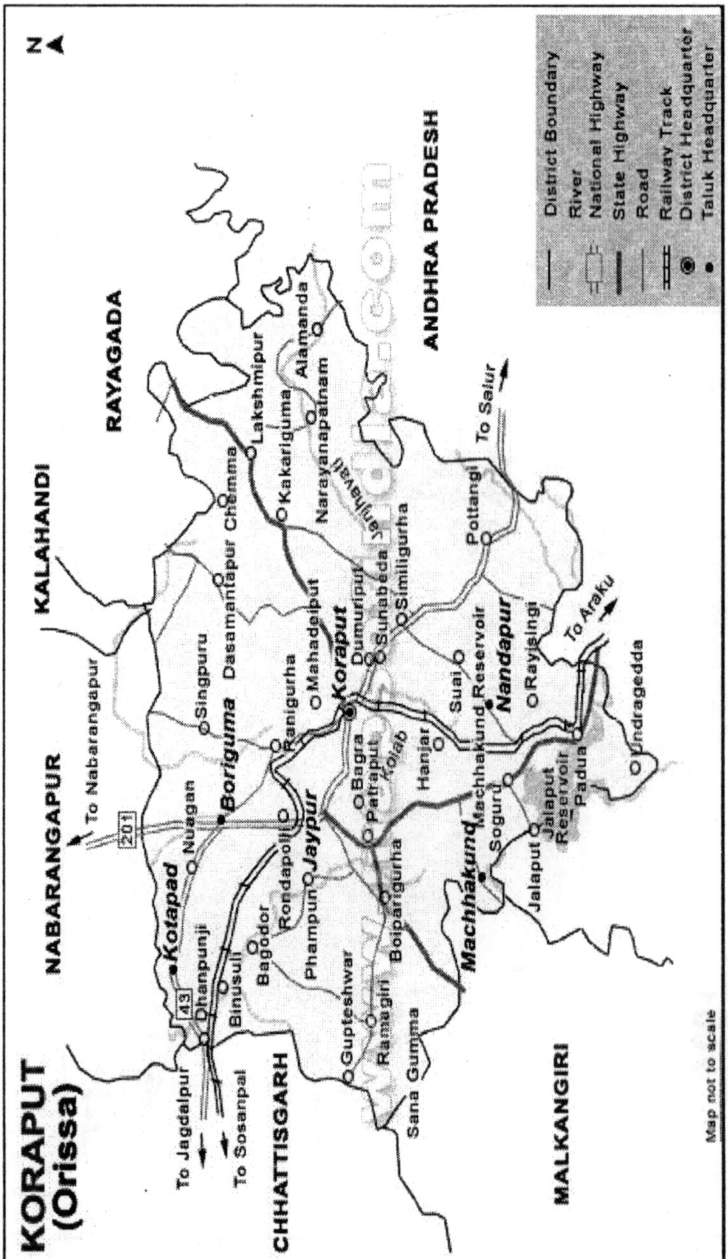

Map II: Koraput District Map.

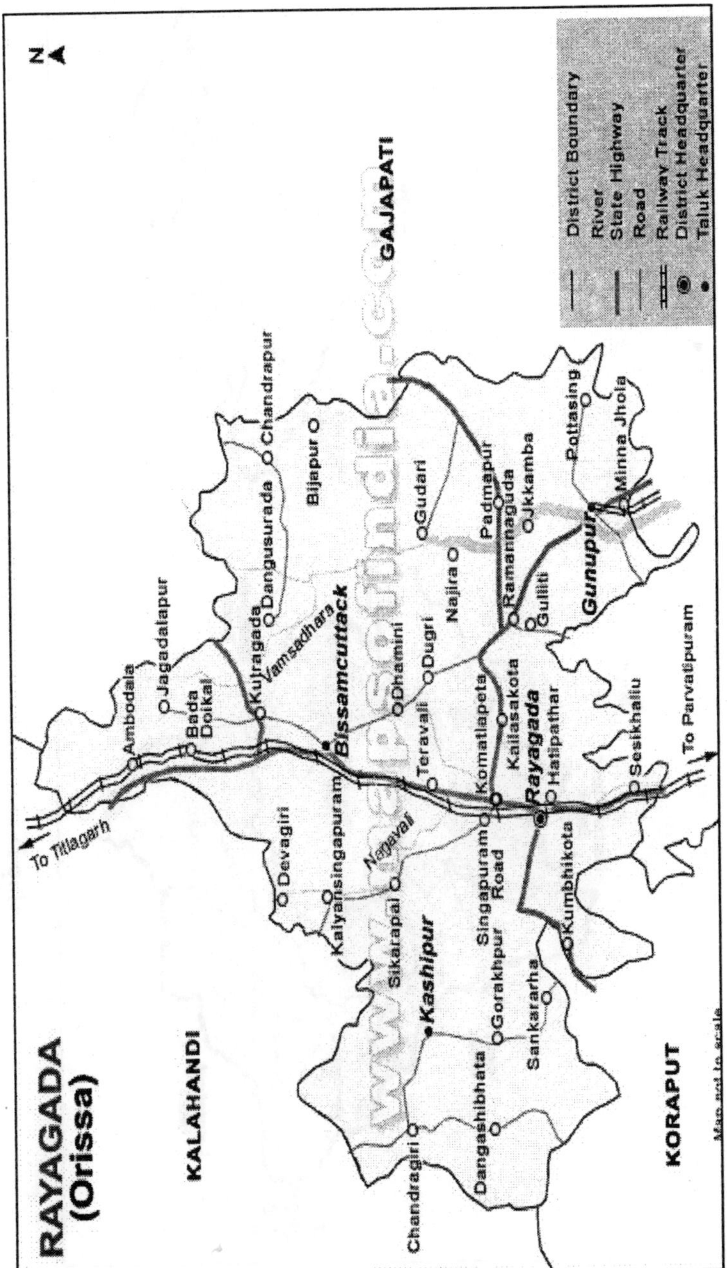

Map III: Rayagada District Map.

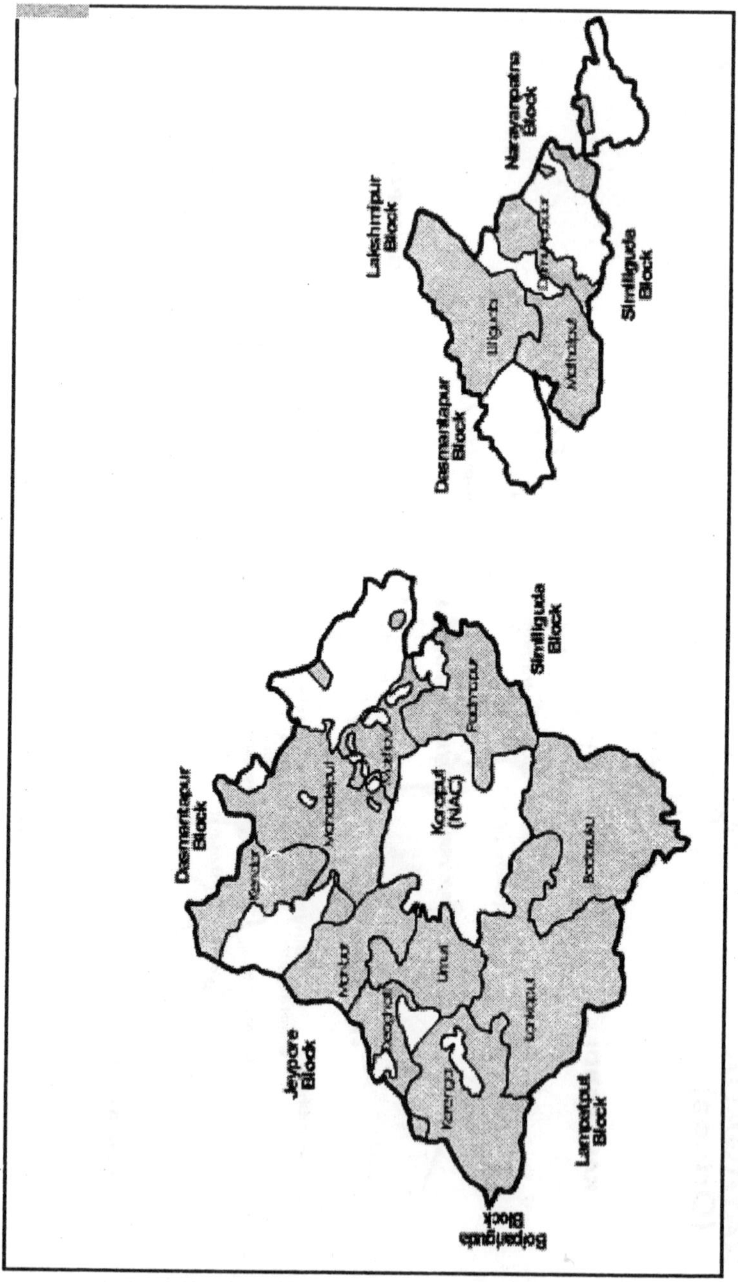

Map IV: Koraput Block Map of Koraput District.

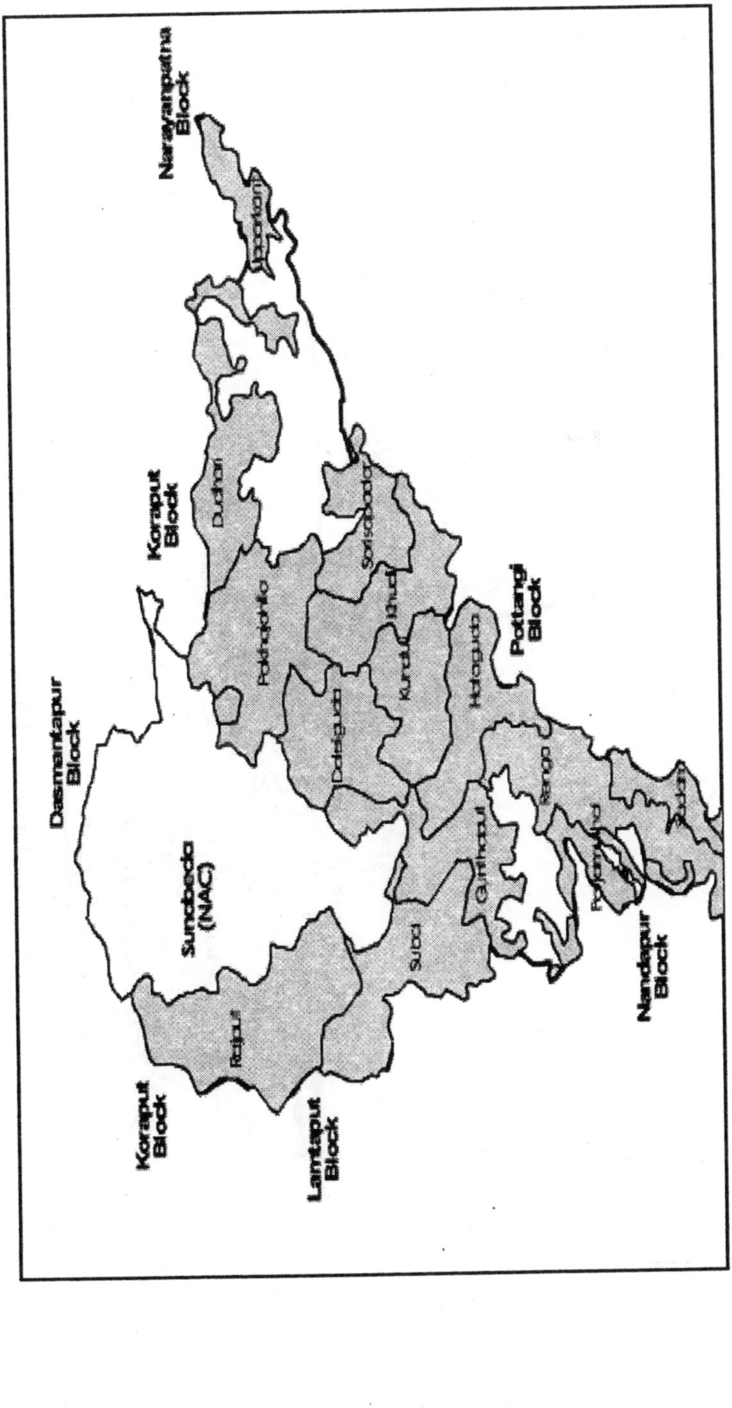

Map V: Similiguda Block Map of Koraput District.

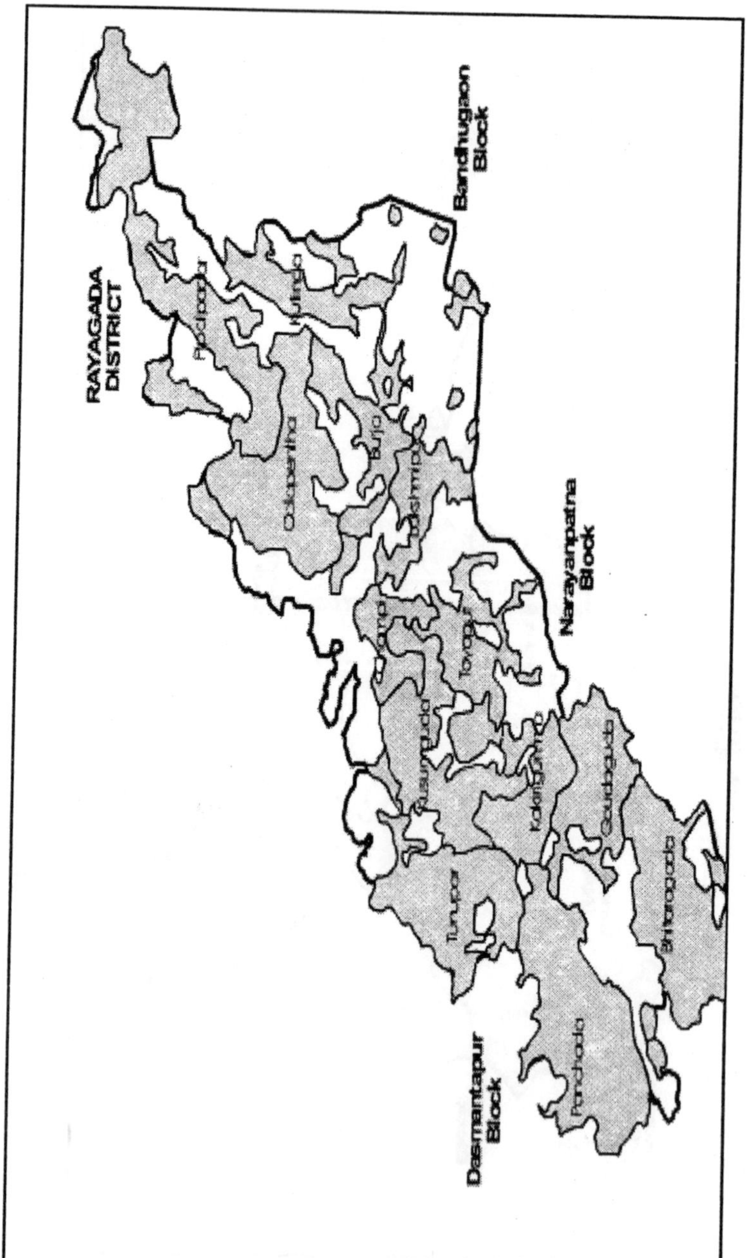

Map VI: Laxmipur Block Map of Koraput District.

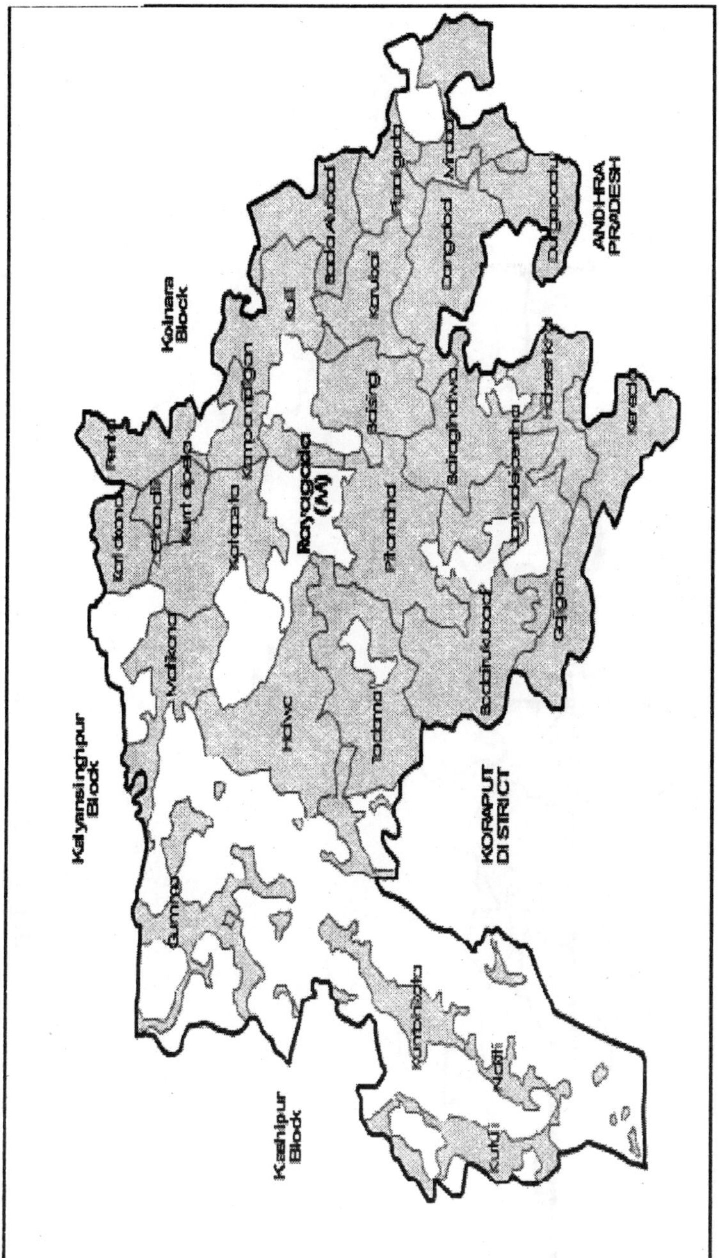

Map VII: Rayagada Block Map of Rayagada District.

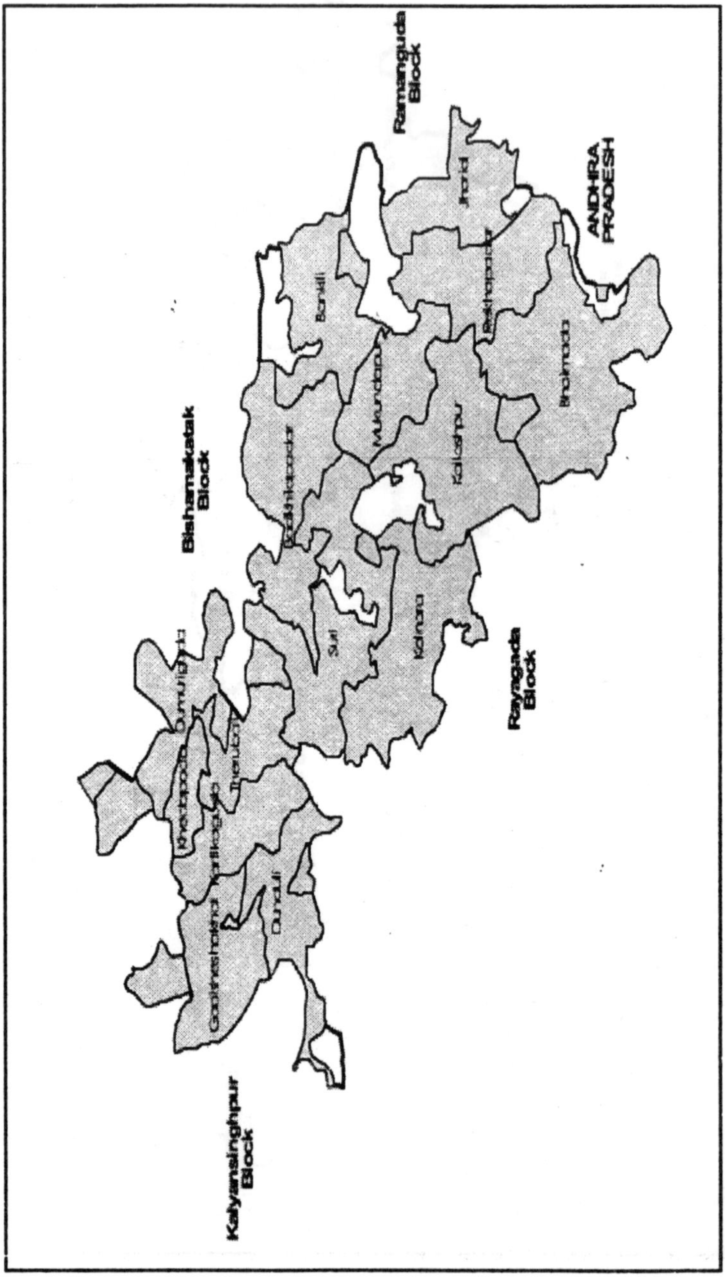

Map VIII: Kolnara Block Map of Rayagada District.

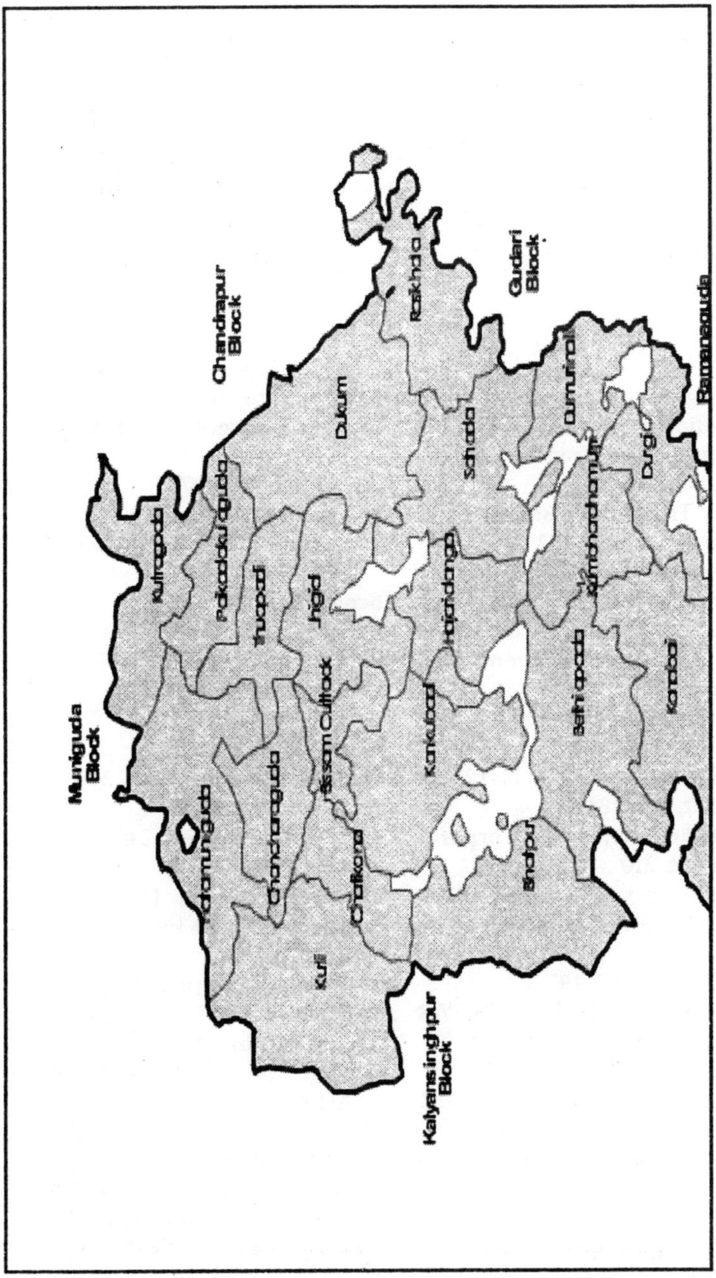

Map IX: Bissamcuttack Block Map of Rayagada District.

Index